SpringerBriefs in Education

We are delighted to announce SpringerBriefs in Education, an innovative product type that combines elements of both journals and books. Briefs present concise summaries of cutting-edge research and practical applications in education. Featuring compact volumes of 50 to 125 pages, the SpringerBriefs in Education allow authors to present their ideas and readers to absorb them with a minimal time investment. Briefs are published as part of Springer's eBook Collection. In addition, Briefs are available for individual print and electronic purchase.

SpringerBriefs in Education cover a broad range of educational fields such as: Science Education, Higher Education, Educational Psychology, Assessment & Evaluation, Language Education, Mathematics Education, Educational Technology, Medical Education and Educational Policy.

SpringerBriefs typically offer an outlet for:

- An introduction to a (sub)field in education summarizing and giving an overview of theories, issues, core concepts and/or key literature in a particular field
- A timely report of state-of-the art analytical techniques and instruments in the field of educational research
- A presentation of core educational concepts
- An overview of a testing and evaluation method
- A snapshot of a hot or emerging topic or policy change
- An in-depth case study
- A literature review
- A report/review study of a survey
- An elaborated thesis

Both solicited and unsolicited manuscripts are considered for publication in the SpringerBriefs in Education series. Potential authors are warmly invited to complete and submit the Briefs Author Proposal form. All projects will be submitted to editorial review by editorial advisors.

SpringerBriefs are characterized by expedited production schedules with the aim for publication 8 to 12 weeks after acceptance and fast, global electronic dissemination through our online platform SpringerLink. The standard concise author contracts guarantee that:

- an individual ISBN is assigned to each manuscript
- each manuscript is copyrighted in the name of the author
- the author retains the right to post the pre-publication version on his/her website or that of his/her institution

Azra Moeed · Stephen Dobson · Sankari Saha

Playful Science Investigations in Early Childhood

A Longitudinal Case Study

 Springer

Azra Moeed [ID]
Te Herenga Waka—Victoria University
of Wellington
Wellington, New Zealand

Stephen Dobson [ID]
School of Education
Central Queensland University
North Rockhampton, QLD, Australia

Sankari Saha
Wellington, New Zealand

ISSN 2211-1921   ISSN 2211-193X   (electronic)
SpringerBriefs in Education
ISBN 978-981-99-7285-2   ISBN 978-981-99-7286-9   (eBook)
https://doi.org/10.1007/978-981-99-7286-9

This Springer imprint is published by the registered company Springer Nature Singapore Pte Ltd.
The registered company address is: 152 Beach Road, #21-01/04 Gateway East, Singapore 189721, Singapore

Paper in this product is recyclable.

Acknowledgements

The authors would like to thank the Faculty of Education, Victoria University of Wellington in New Zealand for supporting this research. We thank the teacher, her students, and her early childhood centres for their support. Thanks to our colleagues who have peer reviewed the work and helped us to improve it. Finally, and importantly, we thank Dr. Abdul Moeed for reading the many drafts and providing critique, guidance, and editorial support.

Abstract

When an early childhood teacher was mentored and supported to create science exploration opportunities, it had positive learning outcomes for the teacher and for the children's science learning. This longitudinal case study was conducted over a period of ten years. Data were gathered through teacher interviews, learning stories written by the teacher, her reflections, collaborative planning notes as well as the mentor's notes and diary entries. Two interviews were conducted each year and over 200 learning stories were first selected that had some science teaching and learning, of these a representative sample of 52 were analysed in depth. Using theories of Communitas, Socio-cultural theory and Social Constructivism the data were analysed both deductively and inductively, triangulated and checked for accuracy.

Mentoring was practiced through semi-structured interviews, phone conversations in a timely manner which were documented, and the mentor being available when there was a curriculum change. The study was a learning journey for both the teacher and the mentor. The mentor evolved from being the fountain of wisdom to a listener and sounding board who enabled the teacher to put forward her ideas and think them through. Mentor meetings became conversations where the joy of children's learning was shared and plans for next steps explored.

Findings suggest that when the teacher was encouraged and supported to try out science related explorations, she gained confidence and both the teacher, and the three- to five-year-old children increased their content and joy of learning. Multiple approaches to science investigation were tried and provided rich opportunities for literacy and numeracy development. Learning stories were used as assessment tools and over time became learning conversations; the former being teacher-centred reporting, to the latter becoming more child-centred and giving the children a voice. We found that learning conversations produced rich assessment information and the teacher became increasingly competent in using this information for future planning. Children were encouraged to talk, think, and say why they thought as they did, thus allowing for metacognitive thinking.

Mentoring that worked in this case was to progressively and with support hand over the teaching and learning responsibility to the teacher. Gradually, the mentor changed her approach from telling and showing to listening and offering well-earned

praise. Other findings were that timely and sustained mentorship was more useful than
~~brief~~ one-day professional development sessions. The teacher learnt and practised
inquiry into her practice and found this rewarding. Children, played, explored, investigated, and learnt science enabling them to make sense of their physical and natural
world from a science perspective. We also found a need for timely and sustained
professional development if the is intention is that an aspirational curriculum is to
lead to aspirational teaching, and rich learning.

Contents

Chapter 1
Playing Teaching and Learning Through Science Investigations

Introduction and Literature Review

Abstract Rich learning experiences provided by the teacher enable children's cognitive development. This chapter presents background information and review of recent selected literature relevant to the multifarious aspects of this longitudinal case study. The focus is upon teacher learning of science ideas, presenting carefully chosen learning experiences in playful contexts and how this relates to children beginning to understand science as one way of understanding the world. Then we address the need for teachers to have the knowledge and pedagogy to present it in a fun and purposeful way. A brief review of literature about play based learning follows and thereafter the focus is upon what is science investigation and what children might learn from investigating in ECE. Also included is an overview of learning stories as assessment tools and how they are used currently and their future potential for assessment, reporting, and planning for learning.

Keywords Science education in early childhood · Approaches to ECE teacher development · Mentoring and coaching · Learning through play · Learning stories

The chapter highlights the context of the study describing curriculum requirements and current practices. This book presents a longitudinal case study of teacher development with the support of a mentor. The chapter ends with information about coaching and mentoring informed by relevant literature. In sum this has been a learning journey of teacher learning and mentoring for both parties.

1.1 Introduction

Children are inquisitive, wanting to explore, observe, try out and fail, then succeed, experience the joy of achievement, and move on to doing the next thing. Human life begins with using our senses to come to understand how the world in which we have arrived, works. Of all animal babies, a human child is born most dependent and takes a long time to learn to feed, walk, talk, and care for themselves. This is not

A. Moeed et al., *Playful Science Investigations in Early Childhood*,
SpringerBriefs in Education, https://doi.org/10.1007/978-981-99-7286-9_1

accidental; the evolutionary processes have ensured that the human baby's head does not become too big to pass through the birth canal, a survival advantage. The smaller skull can only accommodate a smaller brain at birth, but the baby's brain grows quickly after birth. The capacity to develop and learn in the first few years of human life, is exceptional. Herein lies the responsibility of all adults in charge of children, mothers, fathers, other family members and teachers to provide a stimulating, caring, safe, and supportive environment for the children to learn and grow. Born with the single tool of crying, for all purposes the baby must learn everything else. It is up to the adults to provide what Bull calls a 'Library of Experiences' (Bull, 2012, p. 26).

1.2   Science Education in Early Childhood

> Science teaching and learning is about introducing children to the ways scientists think about and investigate their surrounding environment. Scientists do this in two ways.
>
> 1. They explore and confirm ideas about the environment we live in through investigation and exploration.
>
> 2. They form hypotheses or "working theories" to make sense of the surrounding environment and identify these as science knowledge. (Ministry of Education, 2017, n.d.)

Science education in early childhood needs to not only focus on science phenomena but lift its sights to teach what science is and how scientific knowledge is created. Recent research by Hansson et al. (2021) argued that science education in ECE ought to prioritise teaching about the Nature of Science (NOS), how scientific knowledge is created and works. This in line with the *New Zealand Curriculum* (Ministry of Education, 2007) for schools that gives primacy to NOS and place it as an overarching strand, above the contextual strands of physical, material, and living worlds and planet Earth and beyond. Our considered view is that learning about the nature of science ought to begin in early childhood, because teacher understanding of NOS could inform the attitudes and responses of the teacher to children's explorations. This is likely to result in teacher not just passing on the science knowledge but talking about how science works. The intention would then be not to give a response but to encourage children to think why something might happen. For example, when a child drops something, the teacher does not say that the things fall down because of gravity. Instead encouraging children to try dropping a range of objects and coming up with their own working theory about what happens when we drop things.

Young children are inquisitive and when they are asked to do something, for example, go to bed early, or refused a request, their most common question is Why? They know that why is an important question and use it very often. Let's use it to help the children to ask questions and to develop the understanding of their working theories about the observations they make during their exploration of the natural world. We think we can create learning environments where we can ask them, *why do you think this happened*? This is a goal of the Early Childhood curriculum *Te Whāriki* (Ministry of Education, 2017) as illustrated here.

Children experience an environment where:	Over time and with guidance and encouragement, children become increasingly capable of:
• Their play is valued as meaningful learning and the importance of spontaneous play is recognised	• Playing, imagining, inventing, and experimenting
• They learn strategies for active exploration, thinking and reasoning	• Using a range of strategies for reasoning and problem solving
• They develop working theories for making sense of the natural, social, physical, and material worlds	• Making sense of their worlds by generating and refining working theories

Te Whāriki (Ministry of Education, 2017, p. 25)

1.3 Teacher Knowledge and Practice

In New Zealand recent research into ECE science learning has argued that teachers have a significant role "noticing, recognising, and responding to children's scientific interests in authentic and meaningful ways" (Freeman, 2021a, 2021b, p. II). This requires drawing upon their own knowledge of science in their surroundings. ECE teachers have knowledge about children, teaching, learning and curriculum that Edwards and Knight (2000) argue can translate into rich learning experiences. Further that "the teacher must plan learning experiences that engage and challenge children in thinking that is conceptually rich, coherently organised, and persistently knowledge building" (Garbett, 2003, p. 470). At the same time literature suggests a lack of confidence and capabilities with science among ECE teachers (Edwards, 2010). It was suggested that Te Whāriki (1996), the curriculum was non-prescriptive and complex (Hedges & Cullen, 2005). Zhang and Birdsall (2016) argue that few scientific experiences take place in ECE due to a play-based philosophy of practice. Literature has identified a lack of content knowledge within pedagogical practices in New Zealand ECE centres (Hedges, 2021). Furthermore, science learning was supported by teaching strategies impacted by knowledge of science, teacher confidence, beliefs, and attitudes (Edwards, 2010; Freeman, 2021a, 2021b; Zhang & Birdsall, 2016). While some early childhood teachers take a hands-off approach because of low efficacy, others take an unduly instructional approach that is less conducive to play based learning.

Researchers argue that teachers with a broad base of conceptual knowledge of science can recognise science ideas that link with children's interest and present them in purposeful and engaging ways leaving lasting memories of these interactions and experiences (Kumar & Whyte, 2018).

To teach science, teachers need subject matter knowledge (SMK): science content knowledge and knowledge about how science works. The latter includes knowledge of science concepts and procedures, and epistemic knowledge about how science knowledge is created and validated. In addition, teachers need to know how to teach SMK effectively and that is to know how to teach it (pedagogical content knowledge [PCK]).

A large comparative study by Bose and Bäckman (2020) of 64 schools in Botswana and 67 schools in Sweden found that teachers in Sweden had both SMK and PCK, however teachers in Botswana possessed SMK but lacked PCK. Swedish preschool teachers identified science in daily routines and took opportunities to use science and maths. Conversely, teachers from Botswana who lacked PCK found it difficult to teach preschool despite having considerable SMK (Bose & Bäckman, 2020).

Vellopoulou and Papandreou (2019) argue that ECE teachers are not systematically encouraging preschool children's science learning. Further that often, they are not comfortable in teaching content they are not sure of themselves, that is they do not have content knowledge themselves (Kambouri, 2016; Kavalari et al., 2012). Kambouri (2016) argues that teachers do not often ascertain students' prior knowledge or misconceptions and when they do, they do not know how to use this information. Teachers may hold a traditional or idealistic view of children's thinking and perhaps they do not think of them as competent learners (Papandreou & Kalaitzidou, 2019).

Lack of time, science materials or lack of confidence to effectively use the materials is another factor that constrains teachers from implementing science teaching in their classrooms (Nayfeld et al., 2011). Nayfeld et al. (2011) suggest that children's knowledge about science tools and how to use them can be enhanced by engaging them in large group learning experiences for example visiting the rocky shore and being able to explore the rock pools.

Research suggests that how ECE teachers teach science is influenced by their assumptions about young children's capabilities and their own teaching practices (Fleer, 2009). Other factors that may constrain teachers from allowing children to develop scientific ways of thinking could be social, cultural, and institutional (Areljung, 2019). Teachers maintaining the role as leaders in children's learning and transferring scientific knowledge may be a consequence of their traditional pedagogical approaches (Ravanis, 2017). Appleton's (1995) research highlighted that although it has been commonly accepted that teachers believe that a lack of sufficient content knowledge is a barrier to them, they attempted to engage children in science learning that is not entirely correct. Appleton (1995) asserted that factors other than content knowledge such as teacher beliefs and experiences, influenced student teachers' decision to teach science.

Fleer et al. (2021) and other research shows that ECE teachers do not readily engage in teaching science, and it is due to a lack of subject matter knowledge resulting in a lack of confidence. Fleer et al. (2021) propose several reasons why there is a need to investigate models of science teaching in ECE that are research evidence based. They offer several reasons why this is important, among them are; learning takes place in play-based settings; children come to the preschool expecting to play and therefore teachers organise play based activities; teachers are more likely to introduce pre-concepts to do with literacy and the arts and less often with science; and teacher motivation to intentionally teach science also matters.

To focus on the need for more science teaching in ECE Fleer et al. (2021) utilised the Vygotskian concept of crisis. They considered the ECE situation and argue that, on one hand teachers have the institutional demands to introduce innovative pedagogical approaches for science teaching, and on the other hand they have their personal reasons and values. Underpinned by their research Fleer et al. (2021) have presented four theoretical models:

1. Concept crisis – what is the science concept to be taught?

2. Problem crisis – what is the problem to be solved by the children?

3. Crisis of professional practice – how to introduce the concept into children's play?

4. Theoretical problem – what are the relations between play and learning science? (p. 8).

They conclude that the problem of ECE teachers not focussing on teaching science ideas, in not a problem of teacher knowledge, rather it is a problem of practice.

1.4 Teacher Professional Development

In early childhood teaching practice, reflection on practice has been a key aspect of teacher professional development. Cigala et al. (2019) argue that when a group of colleagues reflect on aspects of practice in a systematic and ongoing manner teachers feel more involved in their centre's work processes. A 'thinking group' that is collegial and safe can be helpful in teachers identifying problems and working through solutions collectively.

Quality teaching in early childhood is a significant factor in teacher development. In New Zealand, Clarke et al. (2021) have found it difficult to gain a complete picture of teacher PLD in ECE. Government funded PLD is available through evaluation however, teachers access PLD from a range of providers. Clarke et al. (2021) based on a survey that had 345 responses showed that workshops were the most common and many reported job embedded support but this may not lead to shift in practice. As is the case internationally, most teachers attend one-off workshops and may not experience enabling facilitation strategies identified as useful to shift practice. Clarke et al.'s survey found that 69% of the teachers had attended isolated short-term workshops, 22% experienced a series of related workshops19% said a facilitator visited their centre to support them with their work with the children by observing and giving them feedback. Webinars were experienced by 14% and another 14% said they had attended clustered PLD. The rest, 13% said they had engaged regularly with a facilitator for example, inquiry, communities of practice and action research. The researchers conclude that PLD intended to improve teaching practice needs to focus on the 'who, what, and how' as well as 'why and how often' (p. 79).

Research evidence suggests that mentoring and coaching, which includes observation, providing feedback and engaging in reflective conversations can enhance practice (Dunst, 2015; Kraft et al., 2018; Moreno et al., 2019). Coaching is focussed on specific goals and specified actions and can be effective in helping the teacher with a problem of practice in a timely manner (Elek & Page, 2018). Thornton (2015) suggests that mentoring is more holistic in the context of a sustained professional relationship with long term benefits. Mentoring is an approach that aligns with the findings of the research presented in this book.

Thornton and Cherrington's research into professional learning communities (PLC) found that if the intention was to establish and sustain PLCs in the ECE sector it must have clear membership and effective induction for new teachers, a shared focus, research orientation and commitment. They also recommend having clear goals for leadership, providing opportunities for dialogue and introduction of stimulating new ideas.

Overall, more attention is necessary to ensure that PLD is effective in shifting and strengthening teachers' practice within New Zealand's diverse ECE sector. New Zealand has a variety of contractors and facilitators who provide PLD to the ECE sector, with some PLD programmes funded by the Ministry of Education (MoE) and some contracted privately (self-funded), through early childhood services or umbrella organisations (Cherrington, 2017). Our Code: Our Standards (Education Council, 2017) provides an expectation that all New Zealand qualified and registered teachers will 'engage in professional learning and adaptively apply this learning in practice' (p. 18), and an expectation that teachers' PLD will positively impact on students' learning. While PLD is seen as one of the tools that teachers can use to foster quality teaching practices and improve learning outcomes for children, to achieve this in a diverse workforce it is essential that PLD fits the needs of all teachers (Cherrington & Shuker, 2012).

A range of factors influence teachers' participation choices, including government policy as a driver of the types of PLD available, funding models and cost, teachers' workloads, the availability of relievers for teacher release time, and leadership (Cherrington & Thornton, 2013). Currently Ministry funded programmes include Strengthening Early Learning Opportunities (SELO) (Ministry of Education, 2017), which targets services that support improved participation in quality ECE for Māori, Pasifika, and for children from low socioeconomic families. There are others for example The Incredible Years Teacher (Ministry of Education n.d.-a), Kāhui Ako/ Communities of Learning (Ministry of Education, n.d.-b), PLD to support understanding and implementation of Te Whāriki 2017 (Ministry of Education, n.d.-c, -d), and the Teacher-led Innovation Fund (Ministry of Education, 2019).

1.5 Play-Based Science Teaching and Learning in ECE

Enforced learning will not stay in the mind …Avoid compulsion and let your children's lessons take the form of play' (Plato, The Republic, p. 536).

Research on play-based teaching and learning involves two main areas, play pedagogies and teaching and learning science in early years (Vellopoulou & Papandreou, 2019). In the past, play and learning have been separated when researched for various reasons. In the present times scholars have been critical of such polarisation. Recent efforts have focused on play and how it contributes to children's learning. Much of what needs to be taught in early years is to help children make sense of their physical and natural world that surrounds them. Play is powerful as it provides meanings in everyday activities. It encourages communication, fosters imagination and creativity (Vellopoulou & Papandreou, 2019). Researchers argue that play creates interest, engages the child emotionally and conceptually, allows expression of their growing knowledge and motivates children to investigate and engage in problem-solving (Broström, 2017; Ridgway et al., 2015).

Engaging in play is the most common childhood activity where children encounter science concepts (Bodrova & Leong, 2015; Sikder & Fleer, 2015). Children bring their own knowledge and can learn science ideas in a meaningful way when supported to make links between their everyday experiences and scientific concepts (Fleer & Pramling, 2015). Fleer (2009) suggests that science teaching is possible using free play activities, natural queries, and incidentally. This learning can be an extension of their discoveries, and science learning can be built on their interests, everyday knowledge, and experience by thoughtful teacher's involvement (Fleer, 2009; Larson, 2013).

Sikder and Fleer (2015) in their research proposed four types of *small science*: multiple possibilities for science; discrete science; embedded science and counter intuitive science (see Table 1.1).

Sikder and Fleer (2015) describe the four small science categories.

One- Multiple possibilities are situations where several science ideas are in play in a single activity.

Two- Discrete science concepts occur when activities generally support one line of conceptual development.

Three- Embedded science gives children a scientific lens through which to experience the everyday situation.

Four- Counter intuitive science are those science experiences that are opposite to what is observed.

Sikder and Fleer (2015) provide examples of each of the four types of small science they have reported in Table 1.1. An example of *Multiple possibilities* could be when the child is 'play making a snack'. *Discrete science concepts* can be observed when a child is looking in a mirror. The child may see their image in the mirror and may identify parts of their body. Sikder and Fleer draw attention to the *Embedded science* by explaining that a child experiences the phenomenon of day and night,

Table 1.1 Types of science in everyday activity settings, Sikder and Fleer (2015, p. 460)

Categories of small science	Activity settings	Everyday concepts	Scientific concepts
Multiple possibilities for small science	Preparation of snacks	1. Mixing ingredients 2. Follow the instructions 3. Cooking 4. Concept of shapes	1. Force (push hardly, press, roll) 2. Correlation 3. Properties 4. Change of state of matter 5. Heating and cooling continuum
Discreet science concepts	Mirror play	Identification of body parts	Human body
Embedded science	1. Daytime 2. Nightime (switch on–off) 3. Breathing	Experience of day, night or breathing	1. Light and dark 2. Air 3. Breathing process
Counter intuitive science	1. Sunrise and sunset 2. Moon follows me	Historical development of knowledge	Solar system (earth is moving the position of earth in the universe)

an everyday concept relevant to under 3-year-olds. They suggest that adults who are aware of this can easily help toddlers to learn about the scientific concepts of light and dark. Teaching the abstract concepts in such a way that enables children to experience and everyday situation through a scientific lens. An ideal opportunity for adults to draw attention naming day and night. Adults can communicate to the child darkness is the absence of light. This can be further reinforced when turning off lights at night. Similarly, the abstract idea of air can be introduced by drawing attention to breathing. *Counter intuitive small science* ideas are those that cannot be explained by simple observation. Here the child may well have observed that the moon follows them, which appears to be right, but does not help in learning that the Earth is not at the centre of our solar system. These theoretical understandings through everyday examples are very helpful in introducing science ideas to children (Sikder & Fleer, 2015). Fleer (2021) asserts that "imagination in play is foundational for imagination in conceptual learning, and therefore play-based programs make a key contribution to the development and learning of the young child" (p. 353).

Parker et al. (2022) have described learning through play as joyful, evocative, reiterative, a social, and actively engaging experience. As such it enables fostering cognitive, social, emotional, creative, and physical skills. They found that it was relevant in primary school learning. Parker et al. used a framework for conceptualisation of learning through play for their comprehensive literature analysis. They have found that:

1 Learning through play has a place at school.

2 Playful Pedagogies Can Be Highly Effective.

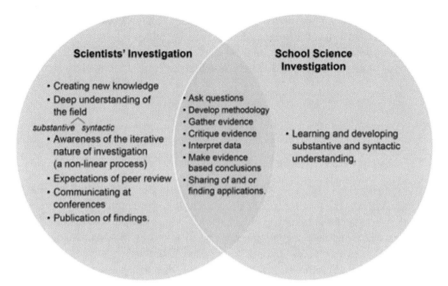

Fig. 1.1 Similarities and differences between scientists' science investigations and school science investigations (Moeed & Anderson, 2018)

3 Effectiveness Is Underpinned by Key Enabling Factors.

4 Effective Playful Pedagogies Combined Facilitation Types.

5 Agency is Central to Playful Pedagogies (p. 751801).

It is possible that when playful pedagogies are used, and children have the agency Parker et al.'s findings may be relevant and useful in preschools.

Zosh et al. (2018) suggest that defining play has been problematic for researchers and scholars. Some describe play as a difficult to access concept, because it is complex, others provide checklists of features of play. Zosh et al. introduce the concept of seeing play as a spectrum. This encapsulates the range from free play, with no guidance or support to guided play where there is adult support, but playful elements are maintained. This novel approach of thinking about play as a spectrum allows for a better understanding of the relationship between play and learning in the context of science learning.

Fleer et al. (2021) and other researchers show that ECE teachers do not readily engage in teaching science, and it is due to a lack of subject matter knowledge resulting in a lack of confidence. Fleer et al. (2021) proposes several reasons why there is a need to investigate models of science teaching that are research evidence based. They highlight the importance of: learning takes place in play-based settings; children come to the preschool expecting to play and therefore teachers organise play based activities; teachers are more likely to introduce pre-concepts to do with literacy and the arts and less often with science; and teacher motivation to intentionally teach science also matters.

Vygotskian concept of crisis. To do this they first argue that, on the one hand teachers have the institutional demands to introduce innovative pedagogical approaches for science teaching, and on the other hand they have their personal reasons and values. Underpinned by their research they offer four theoretical models:

1. Concept crisis – what is the science concept to be taught?

2. Problem crisis – what is the problem to be solved by the children?

3. Crisis of professional practice – how to introduce the concept into children's play?

4. Theoretical problem – what are the relations between play and learning science? (p. 8).

The concept crisis focuses on the science idea rather than the child's playful learning. Whereas the problem crisis puts the child in the centre of the problem for which they are given the agency to play and work out a solution. The notion of crisis of professional practice considers teacher and their professional knowledge or the lack of it and requires teacher learning. Finally, the theoretical problem is to prioritise play, remember that children learn through play rather than presenting the science idea to be confirmed by playing.

They conclude that the problem of ECE teachers not focussing on teaching science ideas, is not a problem of teacher knowledge, rather it is a problem of practice.

An important message from the renowned playwright O. Fred Donaldson:

Children learn as they play. More importantly, in play, children learn how to learn.

1.6   Science Investigation in New Zealand Schools

Internationally, teachers talk about science as a hands-on subject to learn. Often practical work is seen as the best way to learn. There is ample research that suggests that although students find the hands-on tasks engaging, evidence shows that practical work is effective in getting the students to do what the teachers intends them to do, but not always effective in students learning what the teacher intends them to learn (Abrahams & Millar, 2008). However, case study research in New Zealand primary, secondary and Māori medium schools has shown that when teachers do not have too many learning intentions for one practical task and when they share with students what they want them to do *as well as what they want them to learn*, students learn from practical work (Moeed & Anderson, 2018; Moeed & Rofe, 2019). A single learning intention shared with students is more effective than having many learning outcomes from a single practical experience. It is critical in ECE that teachers do not arbitrarily choose something that is of little interest to the children and suggested that early childhood teachers could share a single learning intention with the children.

Practical investigations should not only be *hands-on* but also *minds-on* (Abrahams, 2011). This notion of not just doing but thinking about the doing is we believe an important aspect of investigating in science, irrespective of the level of schooling we are talking about. Hands-on alone is play, a valued way of learning but with play, we might not know what the student is *learning*. Although, in our context of Aotearoa New Zealand, a Māori saying is 'Kei tua o te pae' meaning 'beyond the horizon', suggesting that we can only access a small understanding of young children (Carr, 2009).

Even when we work with very young children it is interesting to find out their working theories about how something works (Te Whāriki, 2017). Then we can support them to develop that science idea further. A minds-on approach takes the activity beyond play to playing and learning science. This book presents research into teacher practice in ECE that helps children to develop their working theories and through trial and error, refine them.

When scientists investigate, they generate new knowledge, however, we note that there is a difference between scientists' investigations and school science investigations. Whereas scientists bring considerable prior knowledge to their investigation and have procedural understanding, children are at best learning how scientists do science and the procedures they follow. Figure 1.1 shows differences and similarities between scientists' investigations and school science investigations.

1.7 Science Investigations in Early Childhood Education

The ECE curriculum Te Whāriki (2017) prioritises exploration and indeed this is one of the most common investigative approaches suggested for ECE. There is agreement amongst early childhood science scholars that very young children can explore, observe, and share their developing science ideas and refine them when supported and guided by an adult (Blake and Howitt, 2012; Freeman, 2021a, 2021b; Kavalari et al., 2012; Schulz, 2012). It is critical to have the opportunity to explore and inquire and to have a supportive adult not telling but encouraging children to try, wonder, and talk to make sense of their sensory experiences. Specifically, Schulz (2012) argues that "many of the epistemic practices essential to and characteristic of scientific inquiry emerge in infancy and early childhood" (p. 382). In her research, Freeman (2021a, 2021b) found that providing a science activity was an invitation to the children and became a provocation when the child engages with the material. Freeman says that a provocation only leads to a meaningful response from the children when the teacher stayed with the activity. She asserts that "The teacher's role was key to facilitating children's engagement with the material in a deep and authentic way through questions, wonderings, and sustained conversations" (p. 33). Earlier, Blake and Howitt (2012) had also highlighted the role of an adult to help children with conceptual understanding suggesting that this role should bring to the fore the everyday nature of science. We found that providing activities that invite, act as provocation and having a teacher on hand to ask questions, comment on wondering and having learning conversations were useful pedagogical practices that supported children's science investigations in our research.

Six types of science investigations have been proposed for school science, that are useful for even for very young children just starting school (Watson et al., 1999). These include:

(1) classifying and identifying,
(2) fair testing,
(3) pattern seeking,
(4) exploring,
(5) investigating models, and
(6) making things or developing systems.

The *New Zealand Curriculum* (Ministry of Education, 2007) calls these types of investigations, 'approaches to investigation' and they are set out as aims within the Nature of Science strand of the science curriculum area.

Investigating in science

• Carry out science investigations using a variety of approaches: classifying and identifying, pattern seeking, exploring, investigating models, fair testing, making things, or developing systems. (NZC, Ministry of Education, n.d.)

We have found that all six approaches to science investigations can successfully be taught in ECE and will provide evidence and examples of each in Chap. 4.

1.8 Early Childhood Education in New Zealand

It has taken about 50 years for ECE in New Zealand to gradually move from provision of care towards an educational function (May, 2019; Freeman, 2021a). Over this extended period, ECE was offered in a variety of settings, meeting the diverse needs of the communities. In New Zealand, ECE provides for children from birth to six years, although most children start school on their fifth birthday. According to May (2019), a range of diverse institutions are providing care, support, and education for these young children. They have different philosophical and cultural traditions, for example Montessori who provide education up to the age of 6-years which is the legal age when children are required to start their schooling. At the last census (2020) there were over four and a half thousand ECE providers, 3786 of these are kindergartens, homebase and education providers and are teacher-led. The rest are parent-led centres and Kōhanga reo.

1.8.1 *Kōhanga Reo*

Kohanga reo are uniquely Māori ECE centre that were established to "stay the decline of te reo Māori me ona tikanga (unique cultural practices), to address issues of sociocultural disruption and concerns of identity loss" (Skerrett White, 2003, p. 8).

Te Kohanga Reo reasserted in a visible way the validity of Māori language, tikanga and akonga. Secondly it provided a focus for the proactive involvement of Māori whānau in the educational development of their children. (Smith, 1989)

Te Kōhanga Reo offered Māori whānau (family) the choice of providing early childhood education which prioritised; Te Reo Māori, their child's first language, living the Māori tikanga, and through interactions with their elders to develop their identity as Māori. The long-term impact was arguing for and having Kura Kaupapa Māori schools which are underpinned by Kaupapa Māori Philosophy. Kaupapa Māori is an ontology and epistemology underpinned by the indigenous knowledge systems of the Māori people (Ormond, 2023, p. 281). In Kohanga Reo, all teaching, learning and interactions are in te reo Māori.

Although ECE is not compulsory, over 90% of 4-year-olds attend an ECE centre.

1.9 *Te Whāriki*—Our Bilingual, World Leading Curriculum

Te Whāriki a te Kōhanga Reo is an indigenous framework immersed in te reo Māori for Te Kōhanga Reo. *Te Whāriki Early Childhood Curriculum* is a bicultural framework for early childhood services. They are not translations of each other. Both have equal status and mana. (Ministry of Education, 2017)

Globally, New Zealand is considered to have a world leading curriculum for two unique reasons; it is bilingual, it integrates education and care at the level of administration, policy, and teacher education (Te One, 2013). In their evaluation of the implementation of *Te Whāriki,* the Education Review Office found that 90% of the 627 centres were using the principles and strands of the curriculum document (Ministry of Education, 2013). However, earlier a scholarly review was critical of Te Whāriki as it did not provide guidance to the teachers about *what and how to teach* (Nuttall, 2013). Consequently, the updated Te Whāriki (Ministry of Education, 2017) has addressed this issue and provided guidance to teachers about what this science learning may look like in ECE.

1.10 Two Approaches to Teacher Development; Coaching and Mentoring

This book presents the learning journey of an early childhood teacher who becomes a confident teacher able to provide rich opportunities for exploration, encouraging children to engage in playful activities that help them to think, solve problems and communicate their growing understandings of the world around them.

In relation to supporting individual's development, two approaches can be taken, coaching and mentoring. Coaching involves the coach clarifying goals, examining current needs, exploring options, agreeing on actions, and implementing and reviewing them. A coaching model is a framework (Priyadharshini & Singaravelu, 2021). The purpose in coaching is to improve, develop, learn new skills, find personal success, set, and achieve goals to manage personal challenges and life changes. Coaching often involves attitudes, behaviours, skills and knowledge, career goals and ambitions, and could include physical and spiritual development (Zentis, 2016). Whereas mentoring is a one-on-one support of a novice or less experienced practitioner (mentee) by a more experienced practitioner (mentor). Mentoring is designed to help the mentee's expertise with the goal of induction of the mentee into the culture of the professions, in this case teaching. Mentoring is in some senses the broader concept and may also involve coaching. The mentor taking on the role of the coach may support the development or one or more job related skills or capabilities (Hopkins-Thompson, 2000). According to Bodoczky et al. (1999) mentors can adopt different roles for example, as an educator, a model to inspire and demonstrate, an acculturate, to induct the mentee into the culture of the profession. Mentor may also provide emotional support and be someone who listens when things are going well or not so well. Templeton et al. (2021) contend:

> Essentially, the mentor mentee relationship requires more than a combined willingness to succeed. Mutual responsibility and respect are foundational pillars, but even more crucial is having a growth mindset and a learning attitude. Mentors who are empowering are also great active listeners, are skilled at powerful questioning, and possess the ability to use self-reflection as a tool for resolute feedback that ultimately builds capacity in others (p. 349).

Mentoring is nurturing, intentional, educational, and supportive action where a more experienced, often older person helps to shape the growth and development of one who is less experienced (Onchwari & Keengwe, 2008). In the context of ECE teacher development mentoring can be an effective way to support teachers to adopt new practices (Weaver, 2004). Professional development practices such as mentoring provide individuals one-to-one guidance, and ongoing support. Collegial mentoring enables the teachers to talk about practice and teach each other what they know about planning, design, and curriculum (Onchwari & Keengwe, 2008).

The purpose of mentoring in this research was to support a non-specialist ECE teacher to gain pedagogic competence in science and not to provide coaching with set goals and helping identify issues, together reflecting on what has occurred and deciding the next steps towards a set goal.

Dewi (2021) has found this approach to be successful with some teachers.

1.11 Learning Stories

Assessment in the New Zealand early childhood education context is required to align with Te Whāriki (Ministry of Education, 1996, 2017). Learning stories were developed as a narrative assessment approach that aligned with the principles that underpinned Te Whāriki (Carr et al., 1998a, 1998b). The "learning stories" had a narrative approach and the purpose was to document children's development and learning. Making them accessible to the children and their families and encouraging them to actively participate in the assessment process were given primacy and active participation of the family was to be encouraged in the assessment process (Cameron, 2022).

The framework proposed by Carr et al. (1998a) required teachers to first "Notice" what children were doing, "Recognise" the learning children were engaging in and to then "Respond" to that learning in the form of next steps for learning. The intention was to bring assessment and planning for learning closer together. Carr also noted that "There were special characteristics of the observations demanded by the learning story framework" (p. 33). She provided guidance on the observation methods to be used that needed to be open-ended and focused and were written in narrative style. Checklists and running records were considered inappropriate and suggested observation methods were to be used. Ministry of Education produced a multimedia resource to support ECE teachers to understand and use learning stories which was led by Carr.

The following is an example of a learning story from Kei tua o te Pae (Ministry of Education, 2004/2007/2009, p. 11). As shown in Fig. 1.2, it exemplifies the teacher noticing, recognising, and responding to the child. It also shows the relationship between assessment and learning by the teacher suggesting adding words to encourage oral language development. However, the example shows that the story is written entirely from the teacher's perspective.

Initially, there was Ministry of Education funding for teacher professional development. Mitchell and Brooking (2007) reported that teachers very quickly adopted learning stories and 78% of ECE teachers were using these by 2003. Further, the learning stories are the dominant type of assessment documentation in ECE.

Hooker (2015) suggests that typically learning stories were placed in the children's learning portfolio where the child's learning journey is documented. These portfolios

---

Blinking and clicking on the changing mat

The teacher (Sue) writes the following observation:

Jace was lying on the changing mat while I was changing him. I was blowing kisses with my mouth.

Jace began to imitate me and do the same action with his mouth.

I then winked at Jace and made a clicking sound with my mouth. Jace once again imitated me and carried out the actions also.

It was really amazing to watch Jace as he looked, listened, and then repeated the actions he saw and heard.

What next?

As well as making facial expressions and sounds, we can add words to what we are doing and encourage more oral language. This can be done throughout all aspects of routines and play.

Fig. 1.2 Example of a learning story

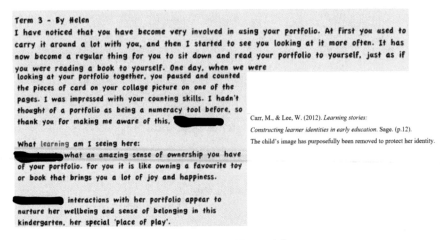

Term 3 - By Helen
I have noticed that you have become very involved in using your portfolio. At first you used to carry it around a lot with you, and then I started to see you looking at it more often. It has now become a regular thing for you to sit down and read your portfolio to yourself, just as if you were reading a book to yourself. One day, when we were looking at your portfolio together, you paused and counted the pieces of card on your collage picture on one of the pages. I was impressed with your counting skills. I hadn't thought of a portfolio as being a numeracy tool before, so thank you for making me aware of this, ▮▮▮▮▮▮

Carr, M., & Lee, W. (2012). *Learning stories: Constructing learner identities in early education.* Sage. (p.12).
The child's image has purposefully been removed to protect her identity.

What learning am I seeing here:
▮▮▮▮▮▮ what an amazing sense of ownership you have of your portfolio. For you it is like owning a favourite toy or book that brings you a lot of joy and happiness.

▮▮▮▮▮▮ interactions with her portfolio appear to nurture her wellbeing and sense of belonging in this kindergarten, her special 'place of play'.

Fig. 1.3 An example of a learning story from a child's portfolio

have photographs along with children's artwork and learning stories. There is a gradual move to e-portfolios that can be accessed by the whānau who may choose to look at them with their child. In Fig. 1.3 we share an example of a learning story that was chosen as an exemplar by Carr et al. (2012).

Brooking (2007) found that 90% of respondents were using photographs to gather assessment information. This suggests teachers' heavy reliance on informal methods. It may be that photographs themselves give authenticity to what the teachers are reporting in their learning stories. There is increasing use of video evidence where the electronic platforms are used to report progress. It was also seen as being accessible to children as children were unlikely to be able to read a narrative by the age of five but can engage with a photograph.

Blaiklock (2008) has raised concerns relating to the reliability and validity of learning stories due to the ways the information reported within them is gathered and the limited quantity of data collected. It appears that teachers would benefit from being aware of the need to gather assessment information via a range of methods, whilst also being knowledgeable about the range of methods available to them, to help strengthen the validity of the data gathered. A related concern is that teachers may be using the learning stories to respond to the children's learning need at that time but perhaps not using it for future planning (Cameron, 2022).

The revised *Te Whāriki* (Ministry of Education, 2017) provides guidance for teachers to further develop their assessment knowledge and practices. Even though the principles, strands, and goals of the original Te Whāriki remained unchanged, the learning outcomes have been reduced from 118 to 20 and more intentional teaching is advocated (McLachlan, 2017). The refreshed version of the curriculum also provides guidance to teachers about assessment highlighting that assessment should take place both formally and informally. The curriculum asks the children to analyse assessment information, document their progress, and use this information for future planning.

Cameron (2022) argues that the assessment of children's learning is complex. Although the learning stories have been developed to specifically align with Te Whāriki, teachers' knowledge of assessment and learning stories seldom reflect the intention of learning stories or guidance given by Te Whāriki. It is important that the teachers have the necessary knowledge and skills to analyse the information they have gathered and to purposefully utilise the analysed assessment information to intentionally support children's future learning. McLachlan (2018) hopefully states "stronger guidance given in Te Whāriki 2017 will help teachers with assessment" (p. 53).

1.12 Summary

This chapter has provided the rationale for science in early childhood education, the importance of teacher knowledge and ability to teach, play-based science learning. We have described what investigating in science means and how are children's investigations similar or different to scientists' investigations. We have presented a brief introduction to ECE in New Zealand and talked about our bicultural ECE curriculum. The chapter presents the background to coaching and mentoring as the approach used for teacher development in this longitudinal research. Finally, learning stories and how they are used for assessment, reporting and future planning is explained.

The findings of this research have shown that change is possible when learning opportunities are made accessible. The Ministry of Education and professional development providers need to consider what time an ECE teacher who has been at the centre from 8 am to 5 pm has to read, understand, practise, and reflect on the increasing educational vocabulary they find on the Ministry of Education website and in professional development sessions.

References

Abrahams, I. (2011). *Practical work in secondary science: A minds-on approach.* A&C Black.

Abrahams, I., & Millar, R. (2008). Does practical work really work? A study of the effectiveness of practical work as a teaching and learning method in school science. *International Journal of Science Education, 30*(14), 1945–1969. https://doi.org/10.1080/09500690701749305

Appleton, K. (1995). Student teachers' confidence to teach science: Is more science knowledge necessary to improve self-confidence? *International Journal of Science Education, 17*(3), 357–369. https://doi.org/10.1080/0950069950170307

Areljung, S. (2019). Why do teachers adopt or resist a pedagogical idea for teaching science in preschool? *International Journal of Early Years Education, 27*(3), 238–253.

Blaiklock, K. E. (2008). A critique of the use of learning stories to assess the learning dispositions of young children. *New Zealand Research in Early Childhood Education, 11*, 77–87.

Blake, E., & Howitt, C. (2012). Science in early learning centres: Satisfying curiosity, guided play or lost opportunities? *Issues and challenges in science education research: Moving forward* (pp. 281–299). Springer, Netherlands.

Brooking, K. (2007). Home-school partnerships. What are they really. *Journal issue*, SET, (3), https://doi.org/10.18296/set.0542.

Bodoczky, C., Malderez, A., Gairns, R., & Williams, M. (1999). *Mentor training: A resource book for trainer-trainers*. Cambridge University Press.

Bodrova, E., & Leong, D. J. (2015). Vygotskian and Post-Vygotskian Views on Children's Play. *American Journal of Play, 7*(3), 371–388.

Bose, K., & Bäckman, K. (2020). Specialised education makes a difference in preschool teachers' knowledge bases in the teaching of mathematics and science: A case of Botswana and Sweden. *South African Journal of Childhood Education, 10*(1), 1–10.

Broström, S. (2017). A dynamic learning concept in early years' education: A possible way to prevent schoolification. *International Journal of Early Years Education, 25*(1), 3–15. https://doi.org/10.1080/09669760.2016.1270196

Bull, A. (2012). Making the most of science learning opportunities in the primary school. *Set: Research Information for Teachers, 1*, 26–28.

Cameron, M. J. (2022). Opportunities and challenges: Assessment in the New Zealand early childhood context. *Assessing and Evaluating Early Childhood Education Systems*, 67–80.

Carr, M. (1998a). *Assessing children's experiences in early childhood. Final report to the Ministry of Education. Part one.* Ministry of Education.

Carr, M. (1998b). *Assessing children's learning in early childhood settings. A professional development programme for discussion and reflection. Support booklet for videos: What to assess? Why assess? How to assess?* New Zealand Council for Educational Research.

Carr, M. (2009). Kei tua o te pae: Assessing learning that reaches beyond the self and beyond the horizon. *Assessment Matters, 1*, 20–47.

Carr, M. (2012). Making a borderland of contested spaces into a meeting place: The relationship from a New Zealand perspective. In *Early childhood and compulsory education*, 92–111. Routledge

Cigala, A., Venturelli, E., & Bassetti, M. (2019). Reflective practice: A method to improve teachers' well-being. A longitudinal training in early childhood education and care centers. *Frontiers in Psychology, 10*, 2574.

Cherrington, S. (2017). Professional learning and development in early childhood education: A shifting landscape of policies and practice. *The New Zealand Annual Review of Education, 22*, 53–65.

Cherrington, S., & Shuker, M. J. (2012). Diversity amongst New Zealand early childhood educators. *New Zealand Journal of Teachers' Work, 9*(2).

Cherrington, S., & Thornton, K. (2013). Continuing professional development in early childhood education in New Zealand. *Early Years, 33*(2), 119–132.

Clarke, L., McLaughlin, T., Aspden, K., & Riley, T. (2021). Supporting teachers' practice through professional learning and development: What's happening in New Zealand early childhood education? *Australasian Journal of Early Childhood, 46*(1), 66–79.

Dewi, I. (2021). A mentoring-coaching to improve teacher pedagogic competence: Action research. *Journal of Education, Teaching and Learning, 6*(1), 1–6.

Dunst, C. J. (2015). Improving the design and implementation of in-service professional development in early childhood intervention. *Infants and Young Children, 28*(3), 210–219. https://doi.org/10.1097/IYC.0000000000000042

Education Council. (2017). *Our code: Our standards: Code of professional responsibility and standards for the teaching profession*. https://teachingcouncil.nz/sites/default/files/Our%20Code%20Our%20Standards%20web20booklet%20FINAL.pdf

Edwards, K. (2010). *The inside story: Early childhood practitioners' perceptions of teaching science*. Unpublished Master's Thesis, Victoria University of Wellington.

Edwards, V., Monaghan, F., & Knight, J. (2000). Books, pictures and conversations: Using bilingual multimedia storybooks to develop language awareness. *Language Awareness, 9*(3), 135–146.

Elek, C., & Page, J. (2018). Critical features of effective coaching for early childhood educators: A review of empirical research literature. *Professional Development in Education, 5257*, 1–19. https://doi.org/10.1080/19415257.2018.1452781

Fleer, M. (2009). Understanding the dialectical relations between everyday concepts and scientific concepts within play-based programs. *Research in Science Education, 39*(2), 281–306.

Fleer, M. (2021). Conceptual playworlds: The role of imagination in play and learning. *Early Years, 41*(4), 353–364. https://doi.org/10.1080/09575146.2018.1549024

Fleer, M., & Pramling, N. (2015). Knowledge construction in early childhood science education. In *A cultural-historical study of children learning science* (pp. 67–93). Springer.

Fleer, M., Fragkiadaki, G., & Rai, P. (2021). Collective imagination as a source of professional practice change: A cultural-historical study of early childhood teacher professional development in the motivated conditions of a Conceptual PlayWorld. *Teaching and Teacher Education, 106*, 103455.

Freeman, S. (2021a). *Opening doors. Guiding teachers to intentionally facilitate science for young children* (Doctoral dissertation, Open Access Victoria University of Wellington| Te Herenga Waka).

Freeman, S. (2021b). Provoking opportunities for science in early childhood education. *Early Childhood Folio, 25*(2), 31–35.

Garbett, D. (2003). Science education in early childhood teacher education: Putting forward a case to enhance student teachers' confidence and competence. *Research in Science Education, 33*(4), 467–481.

Hansson, L., Leden, L., & Thulin, S. (2021). Nature of science in early years science teaching. *European Early Childhood Education Research Journal*, 1–13.

Hedges, H. (2021). What counts and matters in early childhood: Narratives of interests and outcomes. *Journal of early childhood research, 19*(2), 179–194.

Hedges, H., & Cullen, J. (2005). Subject knowledge in early childhood curriculum and pedagogy: Beliefs and practices. *Contemporary issues in early childhood, 6*(1), 66–79.

Hooker, T. (2015). Assessment for learning: A comparative study of paper-based portfolios and online ePortfolios. *Early Childhood Folio, 19*(1), 17–24.

Hopkins-Thompson, P. A. (2000). Colleagues helping colleagues: Mentoring and coaching. *NASSP Bulletin, 84*(617), 29–36.

Kambouri, M. (2016). Investigating early years teachers' understanding and response to children's preconceptions. *European Early Childhood Education Research Journal, 24*(6), 907–927.

Kavalari, P., Kakana, D. M., & Christidou, V. (2012). Contemporary teaching methods and science content knowledge in preschool education: Searching for connections. *Procedia-Social and Behavioral Sciences, 46*, 3649–3654.

Kraft, M. A., Blazar, D., & Hogan, D. (2018). The effect of teacher coaching on instruction and achievement: A meta-analysis of the causal evidence. *Review of educational research, 88*(4), 547–588.

Kumar, K., & Whyte, M. (2018). Interactive science in a sociocultural environment in early childhood. *He Kupu, 5*(3), 20–27.

Larsson, J. (2013). Children's encounters with friction as understood as a phenomenon of emerging science and as "opportunities for learning." *Journal of Research in Childhood Education, 27*(3), 377–392. https://doi.org/10.1080/02568543.2013.796335

May, H. (2019). *Politics in the playground: The world of early childhood education in Aotearoa New Zealand* (3rd ed.). Otago University.

McLachlan, C. (2018). Te Whāriki revisited: How approaches to assessment can make valued learning visible. *He Kupu, 5*(3), 45–56.

McLachlan, C. J. (2017). Not business as usual: Reflections on the 2017 update of Te Whariki. Early Education, *62*, 8–14.

Ministry of Education. (n.d.-a). *Incredible years teacher*. http://pb4l.tki.org.nz/IncredibleYears-Tea cher

Ministry of Education. (n.d.-b). *Communities of learning: Kahui ako*. https://www.education.govt. nz/communities-of-learning/MoE

Ministry of Education. (n.d.-c). *Te Whāriki webinars/Ngākauhaurangi*. https://tewhariki.tki.org.nz/ en/professional-learning-and-development/te-whariki-webinars-nga-kauhaurangi/

Ministry of Education. (n.d.-d). *Te Whāriki professional development workshop materials and videos.* https://tewhariki.tki.org.nz/en/professional-learning-and-development/professional learning-and-development/

Ministry of Education. (1996). *Te Whāriki: He Whāriki Mātauranga mō ngā mokopuna.* Learning Media.

Ministry of Education. (2004/2007/2009). *Kei tua o te pae. Assessment for learning: Early childhood exemplars.* Learning Media.

Ministry of Education. (2007). *He New Zealand curriculum.* Learning Media.

Ministry of Education. (2017). *Te whāriki. He whāriki mātauranga mō ngā mokopuna o Aotearoa: Early childhood curriculum.* Retrieved from https://www.education.govt.nz/assets/Documents/ Early-Childhood/ELS-Te-Whariki-Early-Childhood-Curriculum-ENG-Web.pdf

Ministry of Education. (2019). *Teacher-led innovation fund (TLIF).* http://www.education.govt.nz/ school/people-and-employment/principals-and-teachers/scholarships-for-people-working-in-schools/teacher-led-innovation-fund/

Mitchell, L., & Brooking, K. (2007). *First NZCER national survey of early childhood education services.* New Zealand Council for Educational Research.

Moeed, A., & Anderson, D. (2018). *Learning through school science investigation: Teachers putting research into practice.* Springer.

Moeed, A., & Rofe. (2019). *Learning through school science investigation in an indigenous school. Research into practice.* Springer.

Moreno, A. J., Green, S., Koehn, J., & Sadd, S. (2019). Behind the curtain of early childhood coaching: A multi-method analysis of 5,000 feedback statements. *Journal of Early Childhood Teacher Education, 40*(4), 382–408. https://doi.org/10.1080/10901027.2019.1604454

Nayfeld, I., Brenneman, K., & Gelman, R. (2011). Science in the classroom: Finding a balance between autonomous exploration and teacher-led instruction in preschool settings. *Early Education and Development, 22*(6), 970–988.

Nuttall, J. (2013). Curriculum concepts as cultural tools: Implementing Te Whāriki. In J. Nuttall (Ed.), *Weaving Te Whāriki* (2nd ed., pp. 177–195). NZCER Press.

Onchwari, G., & Keengwe, J. (2008). The impact of a mentor-coaching model on teacher professional development. *Early Childhood Education Journal, 36*, 19–24.

Papandreou, M., & Kalaitzidou, K. (2019). Kindergarten teachers' beliefs and practices towards elicitation in science teaching. *Educational Journal of the University of Patras UNESCO Chair.*

Parker, R., Thomsen, B. S., & Berry, A. (2022). Learning through play at school—A framework for policy and practice. *Frontiers in Education, 7*, 751801. https://doi.org/10.3389/feduc.2022. 751801

Priyadharshini, N., & Singaravelu, G. (2021). Coaching in teaching for effective learning. *International Research Journal of Education and Technology, 3*(2), 40–47.

Ravanis, K. (2017). Early childhood science education: State of the art and perspectives. *Journal of Baltic Science Education, 16*(3), 284.

Ridgway, A., Quiñones, G., & Li, L. (2015). *Early childhood pedagogical play: A cultural-historical interpretation using visual methodology.* Springer.

Ormond, A. (2023). Kaupapa Māori. In *Varieties of Qualitative Research Methods: Selected Contextual Perspectives* (pp. 281-286). Cham: Springer International Publishing.

Schulz, L. (2012). The origins of inquiry: Inductive inference and exploration in early childhood. *Trends in Cognitive Sciences, 16*(7), 382–389.

Skerrett White, M. N. (2003). *Kia mate rā anō a Tama-nui-te-rā: reversing language shift in Kōhanga reo* (Thesis, Doctor of Education (EdD)). The University of Waikato, Hamilton, New Zealand. Retrieved from https://hdl.handle.net/10289/14051

Sikder, S., & Fleer, M. (2015). Small science: Infants and toddlers experiencing science in everyday family life. *Research in Science Education, 45*(3), 445–464. https://doi.org/10.1007/s11165-014-9431-0

Smith, L. T. (1989). Te reo Maori: Maori language and the struggle to survive. *Access, 8*(1), 3–9.

Te One, S. (2013). Te Whāriki: Historical accounts and contemporary influences 1990–2012. *Weaving Te Whāriki: Aotearoa New Zealand's Early Childhood Curriculum Document in Theory and Practice*, 7–34.

Templeton, N. R., Jeong, S., & Pugliese, E. (2021). Editorial overview: Mentoring for targeted growth in professional practice. *Mentoring and Tutoring: Partnership in Learning, 29*(4), 349–352.

Thornton, K. (2015). The impact of mentoring on leadership capacity and professional learning. In C. Murphy & K. Thornton (Eds.), *Mentoring in early childhood education: A compilation of thinking, pedagogy and practice* (pp. 1–13). NZCER.

Vellopoulou, A., & Papandreou, M. (2019). Investigating the teacher's roles for the integration of science learning and play in the kindergarten. *Educational Journal of the University of Patras UNESCO Chair, 6*(1), 186–196.

Watson, R., Goldsworthy, A., & Wood-Robinson, V. (1999). What is not fair with investigations? *School Science Review, 80*(292), 101–106.

Weaver, P. E. (2004). The culture of teaching and mentoring for compliance. *Childhood Education, 80*(5), 258–260.

Zhang, W., & Birdsall, S. (2016). Analysing early childhood educators' science pedagogy through the lens of a pedagogical content knowing framework. *Australasian Journal of Early Childhood, 41*(2), 50–58.

Zentis, N. (posted) (2016). *The 15 types of coaching.* Institute of Organisational Development. nancy.zentis@instituteod.com.

Zosh, J. M., Hirsh-Pasek, K., Hopkins, E. J., Jensen, H., Liu, C., Neale, D., & Whitebread, D. (2018). Accessing the inaccessible: Redefining play as a spectrum. *Frontiers in psychology, 9*, 1124.

Chapter 2
Research Design and Methodology

Abstract This chapter presents the theoretical frame, the research design and methodological approach taken in the present research. The research design is qualitative, and a case study approach is taken. It is an illuminating case study of one ECE teacher's development as a science teacher of preschool children. First the theoretical framework including the underpinning theories is presented succinctly and supported with relevant literature. This leads to the identification of research questions, followed by the research design which is a case study of a teacher's development as a teacher of science supported by a mentor.

Keywords Communitas · Socio-cultural theory · Mentor dairies · Validity and reliability · Researcher reflexivity

2.1 Theoretical Framework

Three theoretical ideas underpin this research, *Communitas, Socio-cultural theory* and *Constructivism.*

2.1.1 Communitas

The notion of communitas is rooted in the Latin concept of communis, meaning common, collective, shared, or possessed by all (Sajewska, 2021). The theoretical concept was developed by Turner (1969), an anthropologist when writing about Liminality and Communitas (Turner et al., 2017). Turner theorized his felt emotions when he explored ritual interactions as he studied different communities in Africa and South America.

Making detailed observations and by participating with communities in their rituals, he put forth communitas as "a collective place of anti-structure, where structure is turned on its head" (Wilmes, 2021, p. 373). Communitas is a place of "spontaneous, immediate, concrete" emotional experience, which he positioned in direct opposition to the more "norm-governed, institutionalized social structures" of many communities (Turner, 1969, p. 372). Wilmes (2021) explains that communitas is a lived experience that occurs when one gets to know and is present in diverse spaces and differs from the concept of community. Community refers to people in a real or virtual place who may not be united through an emotion. Turner's partner elaborated and extended the theory (Turner, 1969). She explains, Communitas "appears unexpectedly in group action. It has to do with the sense felt by a group of people when their life together takes on full meaning." Within communitas one is "freed from the regular structures of life" (Turner, 2012, p. 2) and feel a mutual closeness. Communitas provides a lens on communal life where social structures of daily life are allowed to break, and collective action and emotion is prioritised (Wilmes, 2021).

The theoretical frame of communitas has been adapted by Wilmes (2021) to apply to science learning in an early childhood centre. Using vignettes of interactions within a group of teacher and children she talks about collective joy experienced by the participants engaged in a common learning experience. Their reactions are spontaneous, unstructured, and felt by the members sharing a common experience.

2.1.2 Sociocultural Theory

Te Whāriki the Early Childhood Curriculum is underpinned by sociocultural theory and asserts that "Learning leads development and occurs in relationships with people, places and things, mediated by participation in valued social and cultural activities" (Ministry of Education, 2017, p. 61). Sociocultural theory sits within Cultural-historical theory and is based on Vygotsky's unique concept of zone of proximal development (Daneshfar & Moharami, 2018; Murphy, 2008). Most children (Except for twins and other multiple births) arrive in the world alone. From that point onwards they begin to learn from familial, social, cultural, and interpersonal experiences within their environment. The engagement with cultural connections has an influential impact of the child's social development. We are not born culturally competent member of the society it is essential to be taught the appropriate cultural practices within the interactions in the society (Rogoff, 2003). Some argue that cultural development takes place in two stages: first at the inter-psychological among other people and second, at an intra-psychological level within the child (Veresov, 2017). Veresov proposes the concept of social relations, and that social reality is the foundation of development, and explains that development is the process of the individual becoming a social being.

According to sociocultural theory, learning is a social process shaped by human intelligence in the culture or society the learner lives. Vygotsky's idea highlights the significance of social interaction, where taking part together on a shared activity, even if holding different roles, supports changes in thinking and cognitive functioning. Teaching is seen as the process of helping the learner to develop mental functions. Cognitive theories of learning focus on the mental processes of the individual learner, while sociocultural theories emphasise the involvement of learners in the social practices within their unique context (Danish & Gresalfi, 2018).

2.1.3   Social Constructivism

Social constructivism is one of several constructivist approaches to learning. Constructivism as a theory of learning originated in the field of cognitive sciences and provides a basis for understanding how individuals incorporate new knowledge built from personal experiences into existing knowledge and then make sense of that knowledge (Ferguson, 2007; Tobin, 1990; von Glasersfeld, 2002). Constructivism provides a framework for thinking about the ways in which learners engage and make sense of the objects around them (Bodner et al., 2001; Ferguson, 2007).

The fundamental constructivist belief is that knowledge must be constructed by mental activity of the learner and cannot be transmitted (Driver et al., 1994). Much research into learning in science in New Zealand has been informed by constructivist views of learning since the 1980s (Hipkins et al., 2002) Teaching and learning science draw upon social and personal constructivist views of learning. Understanding which pedagogical approaches are likely to help children grasp concepts, develop skills or understand about the nature of science is likely to influence how we support children's learning (Driver et al., 1994; Leach & Scott, 2003).

2.2   How the Three Theoretical Approaches Informed This Research

Before selecting a learning theory as a lens for this study it was considered important to understand where and how this learning of science investigation was to take place, what is the learning context and what might help us to understand how children were playing and learning science. Early Childhood Centres do not have rigid structures as schools do. In the present research we have also looked at communitas as a collective *joyous feeling*. We felt that the communitas theory best explains the spontaneous joy experienced by the child when they find something new, observe some living things move, see that the large solid ice balloon that they thought would 'definitely drown' does not drown. Equally, when another child says, it is like iceberg. The teachers asks how did you know about iceberg? The child says because I watched Titanic! The feeling that the teacher experiences as a reminder that children bring prior knowledge to the learning situation, can best be explained by the constructivist theory. The joy of learning is then shared by the whole group, in that moment, spontaneously. Our reason for underpinning our research with Communitas.

We see learning in ECE to take place in a social setting where there is interaction between children, adults, and other children. To understand the setting, the participants, the tools for learning and to make sense of cultural and social interactions a social constructivist lens was selected. The pedagogy of the teacher was to develop within this setting through interactions with children, their colleagues and with the children' whānau (Families). Contextually, and for looking at the play, teaching and learning, we selected a sociocultural lens to comprehend our research.

To understand how science knowledge development opportunities facilitate learning, how children scientifically make sense of the play, interactions and exploration constructivist approach was deemed appropriate.

2.3 Research Questions

The main motivation for conducting this research was to understand how a non-specialist ECE teacher can be supported to teach and grow a science teacher. ECE teachers are mostly generalists with few have science qualifications. This we thought was going to require mentor support and encouragement. The working premise was that if the teacher is to focus on exploring science ideas, it ought to result in children's learning. Based on this thinking, the research had the following four research questions:

1. What teacher and mentor interactions support ECE teacher's science teaching pedagogy?
2. How can teacher inquiry into their practice build their science knowledge, knowledge about science, and nature of science investigation?
3. In what ways can science investigations support children to make sense of the physical natural world from a Scientific perspective?
4. In what ways can science investigations be integrated into an ECE learning programme and support children's literacy learning, and their holistic development?

2.4 Methodology

To comprehend teacher development over time, a qualitative approach was considered and within this paradigm, a case study approach was appropriate as the intention was to gain an in depth understanding of the teaching and learning behaviours. The research was conducted as a case study of one teacher over an extended period of ten years. The length of time suggests it is a longitudinal piece of research. However, the length of time is not really what defines the term longitudinal; what is crucial is repeated measurements over a period, which might be as short as a few months. In this case the data were collected yearly over the 10 years of the case study.

Qualitative research is a situated activity; consists of a set of interpretive, material practices that make the world visible; practices turn the world into a series of representations, for example, field notes, interviews, recordings, memos to self, conversations. Importantly, qualitative research studies things in their natural settings, attempt to make sense of, and interpret phenomena from the meanings people bring to the setting (Denzin & Lincoln, 2011, p. 3). A qualitative approach was also appropriate because the research was in Early Childhood Centres. Although classroom observations were not used, data were collected through interviews, mentor, and teacher discussions, field notes, listening to the teacher's point of view, and clarifying meaning, analysis of their reflections and the learning stories documented by the teacher. It was important to keep the mentor and researcher roles separated through being reflexive and reflective. This was achieved by careful recording and transcription of all conversations and interviews. Further, checking of interpretations to make sure they represented the teacher's views by inviting her to read and review. Where there was discrepancy, the meaning made was clarified through speaking with the teacher.

Our decision to take a case study approach was based on further exploration of case study methodological literature. Case studies provide "intensive descriptions of a single unit or bounded system" (Stake, 2006. p. 19). Yin (1994) suggests that case studies are suitable for researching a current phenomenon within its real-life context, "especially when the boundaries between phenomenon and context are not clearly evident" (p. 13). Merriam (1998) defines a case study as an "intensive, holistic description and analysis of a single instance, phenomenon, or social unit" (p. 27). Merriam suggests that case study is appropriate for studying educational practice when the intention is to improve practice. Stake (1995) argues that case study research focuses on an enclosed system which has boundaries and could be the study of a single person, event or system or can be the study of several individual cases studied either concurrently or over time. In this case, our intention was to study a single teacher within the ECE centre where they worked, and this defined the boundaries of this case study.

2.5 Methods

The research was set up originally for two years in an ECE centre where the teacher worked at the time. Ethics approval was sought and gained from Wellington College of Education Ethics Committee. Permission was sought from parents to share the learning stories with the mentor and with the understanding that pseudonyms will be used to protect children's identity.

The fascinating findings about the progress of the teacher led to the agreement to continue with the research for a period of over 10 years. The findings were:

1. A teacher with little or no background in science can with a little timely support become a confident teacher of science in ECE.
2. Teacher confidence grew over time and less support was needed.
3. The teacher learnt to write learning stories that became increasingly detailed and thoughtful.
4. Children were engaging in science and beginning to develop basic science ideas.

As research was to continue it became critical to consider ethical considerations for the safety of all participants. During the research, the teacher worked in three different ECE centres. This meant that informed consent was gained from the head teacher of each of the three ECE centres where the teacher worked during this period. Informed consent was also gained from the parents of the children whose learning stories were shared with the mentor. This included a commitment from the teacher to not share the child's name with the mentor, using only the first letter of their name. As the mentor and researcher at no time visited the centre or interacted with the children, consent was not gained from the children. Ethical practices of not identifying the children were followed by using pseudonyms and not using any pictures of the children in this book.

The mentor's relationship with the mentee was that of a supportive guide and critical friend accessible to the mentee for the duration of this longitudinal research. The mentor had been a science teacher and teacher educator with experience in mentoring preservice teachers and in-service colleagues during her primary and secondary teaching experience.

For this longitudinal research data were collected through:

- Teacher semi-structured in-depth interviews at the start and end of each year.
- Document analysis (teacher planning, reflection).
- Learning stories.
- Teacher inquiry.
- Analysis of shared parent teacher learning conversations and communications.
- Mentor diaries, notes made after each meeting and after each time support provided over the phone. This was an average of about 10 calls over the year in the first 3 years and progressively the frequency of these calls decreased over the years.

2.5.1 Teacher Interviews

For the teacher interviews a guiding protocol was created and followed each year. The first interview was to review the learnings from the previous year and then for the teacher to identify what she had enjoyed, what she had learnt, what were her concerns and to set some goals for the year. The end of year interview was to review the year, what was achieved and what guidance from the mentor would be needed in the following year. The meeting were recorded and transcribed.

At this interview the teacher brought two learning stories that she wanted to discuss. These were discussed and critiqued and ideas about next steps were identified.

2.5.2 Mentor Diaries

Mentor diaries were used to record all interactions some of which were phone conversation or occasional brief meetings. This was critical and very useful in doing the analysis over the years when looking at shift in teacher practice. Advice sought, advice given and any follow up was recorded. The learning stories shared were numbered, dated, and stored. The mentor diary entries comprised 63 pages of handwritten notes that were reflected upon between meetings and analysed.

2.5.3 Teacher Planning

A record of intended teaching, activities designed, issues of finding the time to set these up around the centre's other commitments. This was useful to give the teacher confidence to try things out. Mentor input in planning was useful to ensure that the teacher was developing SMK in living, material, and physical world science ideas. Initially, planning was done together, and in the later years, just run past the mentor for input (Table 2.1).

Table 2.1 Research questions and data sources

Research questions	Data sources
What teacher and mentor interactions support ECE teacher's science teaching pedagogy?	Teacher interviews Mentor dairies Analysis of teacher planning Analysis of learning stories of formative assessment
How can teacher inquiry into their practice build their science knowledge, knowledge about science, and nature of science investigation?	Mentor notes on teacher guidance Planning together Planning independently
In what ways can science investigations be integrated into an ECE learning programme and support children's literacy, numeracy, and their holistic development?	Analysis of learning stories Analysis of teacher–child learning conversations

2.6 Data Analysis

Yin (1994) posits that data analysis consists of examining, categorizing, tabulating, or otherwise recombining the evidence to address the initial propositions of a study (p. 80). Yin also recommends creating a case study database using several sources of data, and meticulously maintaining a chain of evidence, a guidance followed in this instance.

The rationale for using multiple sources of data is the triangulation of evidence. Triangulation increases the reliability of the data and the process of gathering it. In the context of data collection, triangulation serves to substantiate the data gathered from other sources. The cost of using multiple sources was balanced with the manageability of the large amounts of data it created prior to deciding on the use of triangulation.

Yin (1994) suggests that every investigation should have a general analytic strategy to guide the decision about what will be analysed and for what reason. We chose a combination of deductive and inductive analysis. Key aspects of the research would first be analysed deductively using three frameworks. The four strands of the ECE curriculum, salient aspects of science learning, and the identified approaches scientists take to conduct investigations. The following three formed the analysis framework.

1	Strands of Te Whāriki:	Belonging, exploration, communication, and contribution
2	Learning Science:	Conceptual understanding; procedural understanding; and understanding about the Nature of Science
3	Approaches to investigation:	Classifying and identifying, pattern seeking, exploring, investigating models, fair testing, making things or developing systems, The *New Zealand Curriculum*. (Ministry of Education, 2007)

The above three frameworks were used for specific reasons. *Te Whāriki* is the mandated early childhood curriculum. The participating teacher has the responsibility to deliver this curriculum which is intended to lead to the holistic development of the ākonga (Māori word for a learner). As the research set out to investigate teaching and learning through science investigation, our interest was to find out what science ideas were being taught and learnt (Conceptual understanding), what science investigations children engaged in or what science did they do (Procedural understanding). We also wanted to know what the teacher and the children were learning about how science is done (Nature of Science). What scientific skills and behaviours the children are developing, for example, understanding that scientists make observations and offer explanations which are based on evidence. That they look for patterns, they discuss ideas and share these with their peers.

Exploring, making observations, looking for patterns, and communicating their developing theories are all achievable in ECE science learning. This is the stage where children are making sense of the natural and physical world through their learning experiences. The opportunities for science learning in this research were provided by rich learning experiences planned and provided by the teacher.

2.7 Validity and Reliability

Validity is about the credibility of research as believable or not, and reliability is defined as the dependability, and consistency of the method as generating similar results each time the method is used (Cohen et al., 2017). Validity was maintained by using multiple data sets to avoid bias towards one method. Content validity was maintained within the instrument construct (Interview schedule) by asking the same question in more than one way but noticing and recording the differences across the years. The use of multiple tools; teacher semi-structured interviews, teacher reflections, learning stories, teacher planning and mentor diaries allowed for cross-checking of the information collected.

Validity has a cultural aspect, for example, what is valid according to a Māori place-based view may or may not be considered valid in science and vice versa. Paying attention to the cultural and historical environment can be applied for cultural validity balancing against the believable/credibility validity.

Audio recording the interviews and transcribing them allowed for going back and verifying anything that was unclear. The synergy between, the planned activity, teacher reflection and the document learning stories provided confidence in the interpretations made throughout the research journey. This process raised the reliability of the findings.

2.8 Researcher Reflexivity

Reflexivity was an essential feature in this qualitative research as the mentor had two roles, that of the mentor and the researcher (Creswell, 2015). This meant careful consideration of the phenomenon being investigated and the research process which requires what Creswell calls good data (Watt, 2007). The mentor had to take care to constantly reflect on her own assumptions and behaviours as a researcher for reflexivity. The mentor transcribed and analysed the transcripts on an ongoing basis. This was helpful in going back and checking anything that was unclear. Analysis of data each year also made it possible to check the interview comments aligned with the planning and learning stories. This method of analysis helped discuss and decide the next goals. The process of reflecting on the dual roles of mentor and researcher was a sustained and iterative process of reflection and critically analysing the data from the dual perspectives. As the relationship between the teacher and mentor grew, extra care had to be taken to stick to what the data was telling rather than what the mentor/researcher thought. This process was time consuming but was necessary for the validity of our research.

There was also an awareness that the mentor needed to keep abreast with the changes in ECE policy and curriculum over a 10-year period.

Having been involved in science research in a Kura Kaupapa Māori for six years, it was our practice to ask the principal of the school, a very experienced and knowledgeable kuia (elder) to check our understanding of the Māori concepts used. We are very aware that not being Māori we can only offer an outsider's view of the Māori knowledge and worldview.

References

Bodner, G., Klobuchar, M., & Geelan, D. (2001). The many forms of constructivism. *Journal of Chemical Education, 78*, 1107.

Cohen, L., Manion, L., & Morrison, K. (2017). *Research methods in education.* Routledge.

Creswell, J. W. (2015). *A concise introduction to mixed methods research.* Sage.

Daneshfar, S., & Moharami, M. (2018). Dynamic assessment in Vygotsky's sociocultural theory: Origins and main concepts. *Journal of Language Teaching and Research, 9*(3), 600–607.

Danish, J. A., & Gresalfi, M. (2018). Cognitive and sociocultural perspective on learning: Tensions and synergy in the learning sciences. In F. Fischer, C. E. Hmelo-Silver, S. R. Goldman, & P. Reimann (Eds.), *International handbook of the learning sciences* (pp. 34–43).

Denzin, N. K., & Lincoln, Y. S. (Eds.). (2011). *The Sage handbook of qualitative research.* Sage.

Driver, R., Asoko, H., Leach, J., Scott, P., & Mortimer, E. (1994). Constructing scientific knowledge in the classroom. *Educational Researcher, 23*(7), 5–12.

Ferguson, R. L. (2007). Constructivism and social constructivism. *Theoretical Frameworks for Research in Chemistry/Science Education*, 28–49.

Hipkins, R., Bolstad, R., Baker, R., Jones, A., Barker, M., Bell, B., Coll, R., Cooper, B., Forret, M., Harlow, A., Taylor, I., France, B., & Haigh, M. (2002). *Curriculum, learning and effective pedagogy: A literature review in science education* (pp. 70–180). Wellington.

Leach, J., & Scott, P. (2003). Individual and sociocultural views of learning in science education. *Science and Education, 12*(1), 91–113.

Merriam, S. B. (1998). *Qualitative research and case study applications in education* (2nd ed.). Jossey-Bass Publishers.

Ministry of Education. (2007). *The New Zealand curriculum.* Learning Media.

Ministry of Education. (2017). *Te Whāriki. He whāriki Mātauranga mō ngā mokopuna o Aotearoa: Early childhood curriculum.* Retrieved from https://www.education.govt.nz/assets/Documents/Early-Childhood/ELS-Te-Whariki-Early-Childhood-Curriculum-ENG-Web.pdffrom   https://www.education.govt.nz/assets/Documents/Early-Childhood/ELS-Te-Whariki-Early-Childhood-Curriculum-ENG-Web.pdf

Murphy, R. (2008). Dynamic assessment precursors: Soviet ideology and Vygotsky. *The Irish Journal of Psychology, 29*(3–4), 193–233.

Rogoff, B. (2003). *The cultural nature of human development.* Oxford University Press.

Sajewska, D. (2021). Toward theatrical communitas. *Pamiętnik Teatralny, 70*(3), 15–56.

Stake, R. E. (1995). *The art of case study research.* Sage.

Stake, R. E. (2006). *Multiple case study analysis.* Guilford.

Tobin, K. (1990). Changing metaphors and beliefs: A master switch for teaching? *Theory into Practice, 29*(2), 122–127.

Turner, V. (1969). The Ritual Process: Structure and Anti-Structure. New York: Aldine de Gruyter.

Turner, E. (2012). *Communitas: The anthropology of collective joy.* Palgrave McMillan. https://doi. org/10.1111/anhu.12009

Turner, V., Abrahams, R. D., & Harris, A. (2017). *The ritual process: Structure and anti-structure.* Routledge.

Veresov, N. (2017). The concept of perezhivanie in cultural-historical theory: Content and contexts. In *Perezhivanie, emotions and subjectivity* (pp. 47–70). Springer.

von Glasersfeld, E. (2002). Problems of constructivism. In *Radical constructivism in action* (pp. 19–25). Routledge.

Watt, D. (2007). On becoming a qualitative researcher: The value of reflexivity. *Qualitative Report, 12*(1), 82–101.

Wilmes, S. E. (2021). Interaction rituals, emotions, and early childhood science: Digital microscopes and collective joy in a multilingual classroom. *Cultural Studies of Science Education, 16*(2), 373–385.

Yin, R. K. (1994). Discovering the future of the case study. Method in evaluation research. *Evaluation Practice, 15*(3), 283–290.

Chapter 3
Teaching Science in Early Childhood

A Case Study of Teacher Development

Abstract This chapter presents a mentor supported, ten-year learning journey of an ECE teacher from pre-service to in service as a professional. The teacher was a migrant to New Zealand, she had trained and taught in her home country, India and has brought up a son. As Aotearoa, New Zealand was very different and schooling here was totally different to her experience, she made the difficult decision to train to be an ECE teacher and completed a degree in early childhood education. The three-year degree provided the opportunity to develop her English and experience a range of ECE centres and their approaches during this time. She is both a thinker and a doer and turns out that she cares about the children and their learning. This chapter focusses on the professional development of this early childhood teacher who was supported by her mentor to become a competent teacher of children learning through play and make sense of their play with her guidance. Finally, an analysis of the mentoring as documented in mentor diaries and teacher interviews is presented. The chapter concludes with a summary.

Keywords Case study of teacher development over 10 years · Teaching Science in ECE · Examples of science investigations · Teaching as inquiry

3.1 Introduction

In the first two years of ECE teaching this teacher Seema (pseudonym) followed the routines of her kindergarten. Here we present the evidence of Seema's development from a novice ECE teacher to a confident science teacher. The first author has been the mentor and the researcher during this extended period of the teacher's development. The data shared here is from the analysis of:

- Twenty audio-recorded and transcribed interviews.
- Seema's written reflections (approximately 40 pages).
- Mentor's diary notes (63 handwritten pages) and
- Fifty-two learning stories.

The learning stories are purposefully selected from over 220 stories written by Seema and shared with the researchers. The main criteria for selection of stories was that a story demonstrated an aspect of Seema's development as an ECE teacher, and as a science teacher of little children. All names of children and those of the teacher's colleagues are pseudonyms; care has been taken not to identify the ECE centres where the teacher has worked in her teaching journey.

When we began this research project Seema was in her third year as an ECE teacher. She loves children and according to her head teacher at the time:

> You listens and learn, are very aware of safety and care of children. The children love you and enjoy working with you. You are always here early to set up and help your colleagues (Head teacher comment from the end of year meeting).

As her mentor, I had known Seema and was curious and wanted to know what she does with the children. She described her day:

> I go to the centre at 7:30 and set up inside. Put out paint, paper on the easels, set up the collage table. We have the dress up corner, blocks, paper, and crayons for drawing. Outside I take the cover off the sandpit and bring out the sandpit toys, diggers, digging toys and buckets. If it is a fine day, we put out the water trough and have some toys for the children. We also have bikes and outdoor play equipment. Sometimes I read to the children, and we also sing. We take the children for walks.

When asked how children's learning was documented and shared it appears that like other teachers' she wrote learning stories. Here is an example:

> Sam, I noticed that when you walked in this morning you were happy and said good morning to me. Then you went looking for you friend to play with in the sandpit. You and Kyle were very busy shifting the sand from one place to another. You were using the front-end loader to pick up the sand and the tip truck to move it. I think you were making a castle.

Seema said that they paste the learning story into the child's profile book. These books are available for the whānau and children to have a look at any time.

> I gained an insight into learning stories. Will be thinking more about these. They appear to be a teacher's account of what they have seen the child 'do'. Noticing is great but what do they do with this information. I am wondering what it means for the children to read their profile book. It appears all that they can do is look at the pictures. Great reminder but how can this be written from the child's perspective? Is there a place for teacher talk? (Mentor meeting notes to follow up at the next meeting, Year 1of the study).

3.2 From Fear to Fascination of Teaching Science

In this section we share selected learning stories and their comparison across the years.

At the beginning of this research, when asked if she did any science with the children, her response was "I don't know science" how can I teach science to little children.

Not having a background in science Seema does not think she knows much science. I suggested that children could learn about fruits. They could use their senses to see, feel, smell, and taste. Seema thought this was a good idea and she could ask questions to encourage children to explore and talk about the fruit. She decided on apples, and I think this was because most children would be familiar with this fruit (Mentor notes, Year 1).

Seema bought red and green apples for her first science activity. This was something she knew about and could see herself talking with children so that they would learn something about fruit. With some apprehension and excitement, she thought that would be doing a *science* activity. Her development and progression in learning is presented in the next three vignettes which is followed by analysis.

Fruit of the week

Vignette one

We decided to have a focus on one fruit each week. Today's fruit was apple. The children all looked at the apple and said it was round, red, and green. When we cut it up and they tasted it, Samara said it was sweet and Tim said it is crunchy. Children enjoyed eating it. We talked that it can be different colours, red or green. Apples are also juicy. Leah said she likes apple juice (Year 1 story).

Yummy pear, our food of the week!

Vignette two

When possible, we choose a seasonal fruit and this week pear was our fruit explore. Susan, Mili, Sara, Tom, Nathan, and Rangi sat around the table. First, we talked about pears and then children were given the opportunity to take a closer look, feel it, describe its shape, and talk about how its skin felt. Susan said, it is not round. Tom added it is round, but it has a long bit on top. After some thinking, Tom said that the shape of the pear was like a spaceship, while Sara described it as a circle. Mili described it as a triangle and Nathan said its shape was like a giant belly button. Rangi waited his turn and got the shape sorted for us; he said it is pear shaped! Indeed, what better way to describe a pear? We talked about a new shape that we had not learnt yet. The pear was like a cone.

Next, we cut the fruit in half, we looked at the inside and saw the seeds. Then we all ate a piece of pear and children talked about the fruit being juicy, how it smelt and tasted. Then they drew pictures of the fruit (Year 2).

Teacher reflection

Exploring fruits and vegetables became a weekly activity. It is intended to familiarise children with fresh fruit and vegetables, get them to taste these with the long-term goal of encouraging eating fresh fruit and vegetable. This activity provides an ideal opportunity for communicating their ideas. It also provides wider insights, for example, some children chose mathematical shapes to describe the fruit, triangle, and circle. It was a good opportunity for formative assessment. It shows us that our children are remembering the names of the different shapes and are being able to apply it is a real context. Perhaps in future we could think about a part of our body that looks like the fruit or vegetable we are exploring. Children are interested in their bodies and that provides opportunity to extend Nathan's interest.

(Year 2 story)

Making fruit salad

Vignette three

> In term one, children have learnt about different fruits. In first few weeks we did stone fruits because it was a seasonal fruit. Children compared, how each felt, smelt, and tasted. They learnt that they all has a stone as a seed. Nessy said, "plums were red, sweet, very juicy, she loved them, and she asked if she could have the whole plum". Others also got to eat one each. Most children could recognise apricots and peaches, but no one knew about nectarines. That was our new learning. There was lots of opportunity to learn words like the shapes, colours, and textures.
>
> Later, we did pip fruits, apples, pears, oranges, and lemon. Children found lemons were sour. It was a good opportunity to learn about shapes, how each feel, did they need to peel the fruit. Tasting them and getting to eat the fruit was their favourite part. We also learnt about soft and squishy bananas. Nola said it was the shape of a moon, and Sam said that its colour was kōwhai. Ka pai Sam, for remembering the te reo Māori name of yellow.
>
> We learnt about berries, strawberries, and their tiny seeds. Blue berries that were sweet and our cook helpfully made muffins with them for morning tea. We also got blackberries and children said they loved thè taste which was a "little bit sour and a lot sweet" according to Duncan.
>
> Today was my planned day for assessment and to make it fun I decided that the children will make fruit salad! We had apples, pears, oranges, feijoa, banana, and mandarins. Six children were present. Sam talked about the shape of each fruit. Sara corrected that the orange was round, but mandarin was a bit flat. Ness added that it was sort of oval. Tim said the feijoa was oval. They described the textures as crunchy, soft, juicy, and squishy. Then we played the smell and taste game. One child would close their nose and eyes, and another would pick up a piece of fruit and with a toothpick and put it in their mouth. They had to say what fruit it was. We played this game for a long time. Children could explain that if you can't smell then it is the texture and juiciness in the mouth that helped to identify them. Tane said that the oranges were the juiciest, it is his favourite. We had a lot of fun. In the end Fiona who had been very quiet, added, "we did not have any stone fruits"! That is right Fi, the stone fruits are still growing and not ripe yet. Duncan said we had peaches and ice-cream for tea and how mum had them out of a tin (Year 6).

Teacher reflection

It was a long activity today, from morning tea to lunch time. All children stayed which shows they were interested. Getting to eat the fruit also helped. They used describing words for shapes, how they felt and colours. We now know our colours in English and Te reo Māori. Using their sense of taste, they identified the fruits. They found it easy when they could also smell the fruit. Towards the end some would slightly open their eyes and others would tell them off!

Fruit of the week has encouraged all children in their language development. It is a useful sensory activity that makes children use all their senses. Our children will eat any fruit that we have had this year and enjoy it. Using it as an assessment opportunity was good because we have evidence of these children's learning. A few children have left for school, but they had learnt a lot already. Children are learning to compare and say why two fruits are the same or different. Sam said banana is long, soft, yellow and apple is round, red, crunchy. We must peel the banana. Usually, Sam is quiet and does not say much (Year 6 story).

3.3   Analysis of the Above Three Teaching and Learning Experiences

The idea of fruit of the week was suggested by the mentor. As can be seen in *Vignette one, teacher's* presentation of the activity led to engagement. They were able to describe the colour, shape and taste and there was little probing from the teacher. However, she gained confidence to make this fruit of the week a routine for the rest of the term. *Vignette two* provides an interesting insight that she had been teaching shapes and that she took the opportunity to introduce another shape. The other notice-able change was her beginning to write her reflection about the activities she was doing with the children. She was being able to identify Nathan's interest in his own body and made a note to use the body as a context for future activity. However, even at this point she is not thinking about the human body having anything to do with science!

Vignette three shows that with experience, the teacher must consider it a worth-while learning activity and one that has continued in subsequent years. She is now thinking about what the children are learning through this activity. Complexity increases and we see the differentiation of fruit into stone and pip fruits and berries. Perhaps the most significant change in practice is the consideration of assessment. So here is a confident teacher, who knows what science she wants the children to learn, plans the activity, presents it in a fun and engaging game and is looking for evidence of learning. There is also an awareness of higher order thinking, the children are not just describing but also comparing the similarities and differences. She is now confident to use te reo Māori and children are beginning to learn colours and some phrases of te reo Māori.

In sum, the investigation is still exploration, but children are being encouraged to look for patterns and being asked to compare. The pedagogy makes sure that all senses are being used and children are sharing their new learning which is based on observational evidence.

3.4 Analysis of Fruit of the Week Activity Over the Period of the Research

Seema has conducted this activity each year and there are 22 learning stories that describe these. A summary is provided in Table 3.1.

Year	Activity	Teacher practice
1–3	Exploring fruits	Began with apples and added common fruits. One fruit each week. Fruits included apples, bananas, pears, and oranges Other than sensory exploration, talk moved to things that needed to be peeled and those that did not
4–5	Fruit of the week	The activity became a routine. First for a term and then repeated in the last term. Reason, some children had transitioned to school and new ones had arrived She made it a habit to write her reflection. These were sometimes shared with the mentor, along with the stories
6	Seasonal fruits	The idea of using fruits that were in season began in the sixth year. She introduced mandarins, and children's language development was considered. Other than colour and shape, children were talking about juicy, squashy, and crunchy fruits More questions were being asked and as the confidence has grown so had the complexity. She is assessing learning and reporting on it

3.5 Teaching as Inquiry

There are many examples of professional development through teacher inquiry in ECE. For example (Handscomb & MacBeath, 2006), spirals of inquiry (Halbert & Kaser, 2013); teacher professional development through inquiry. Reflecting on practise is encouraged in ECE setting. However, Cherrington and Thornton (2013) posit the need to balance individual reflection with collaborative learning and shared critical reflection.

The New Zealand Curriculum (NZC) (Ministry of Education, 2007) requires teachers to inquire into their practice with the aim of enhancing student's learning. There is a long tradition of inquiry-based teacher education which seeks to transcend the view that it can take place within the artificialness of the university lecture hall or tutorial (Lammert, 2020).

Our reason for using Teaching as inquiry was that the teacher was provided with a photocopy of this framework and had to work out how this inquiry was to be conducted. It is a model for school years in the New Zealand Curriculum. It is also important to note that during this research the teacher was provided with at least six different templates to conduct her inquiry which ranged from a two-page template to one that had 10 pages. In each case she was told to choose something she needed to improve and follow the template. At the end of the year there was a discussion between the teacher and the head teachers mostly to check that the due process was followed. The example we have used here was relevant because the teacher inquiry focus had been on science.

The following section details teaching as inquiry and provides the framework suggested in the New Zealand Curriculum (Ministry of Education, 2007). This is followed by how the teacher in this research inquired into her science teaching practice with a view of enhancing children's learning.

3.5.1 Growing as a Science Teacher of Little Children

Teaching as inquiry provides a framework for the teacher to first finding out what are children's learning needs, and what they want to focus upon. Then the next stage is to plan their inquiry considering what evidence-based strategies are most likely to help the children learn. Gathering evidence of children's learning is undertaken and finally, evaluating what happened due to the change in teaching and learning (Fig. 3.1). Teaching as inquiry is "not a 'project' or an 'innovation' but a professional way of being" (Timperley et al., 2014, p. 22). Teaching as inquiry is a teacher exploration of the impact of teaching and the relationship between teaching and learning. NZC suggests a framework (see Fig. 3.1).

Early childhood teachers are being encouraged to inquire into aspect of their teaching for improving their teaching children's learning. Different Early Childhood

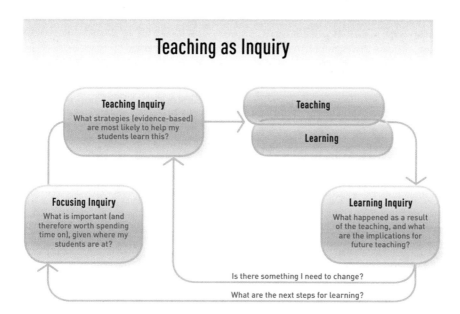

Fig. 3.1 Teaching as inquiry model (Ministry of Education, 2007, p. 35)

Centres (ECE) have come up with their own cycles of inquiry and although they are presented in different forms the focus is still on focussing on one's teaching and noticing its impact on children's learning. This they are doing by taking action, gathering evidence of learning to evaluate and if required improving future practice. Some centres are making such an inquiry part of their teacher professional development and including it in their yearly appraisal cycle. This was the case in the centre where the inquiry shared here was conducted. The ECE curriculum (Ministry of Education, 1996, 2017) has five strands: Wellbeing, Belonging, Contributing, Communicating and Exploration. The teachers can choose an aspect of their teaching they would like to focus upon and improve their practice. In the study, practice at the ECE centre was for the teacher and the head teacher to meet and collectively decide what might be an area for the teacher inquiry. This appeared to have two benefits, one that the teacher must focus, consider, and arrive at the meeting with their thoughts about what they want to inquire. The other that the head teacher who is looking at the bigger picture can help guide and align the inquiry with the current goals of the centre.

3.6   Science Teaching Inquiry Cycle

Seema prepared for her meeting with the head teacher having decided that she wanted to try and improve her practice of teaching science to her children. She had considered the current situation as set out in the Noticing section of the inquiry cycle in Fig. 3.2. After a meeting with the head teacher, she came away happy after collectively deciding on a research question to explore, which was:

How can I and my colleagues support our children to carry out science investigations through setting up exploration opportunities?

The question is crafted in such a way that Seema could work on teaching science investigation in her centre. From the organisation's perspective, she would be focussing on exploration which is the curriculum strand her centre would be concentrating on during the inquiry period. Seema's project widened the possibility of including other teachers who may also want to provide opportunities for exploration in a science context. The inquiry process followed by Seema is explained in Fig. 3.2.

The above inquiry process shows thoughtful planning, resourcing, and inclusion of others.

3.6.1   Evidence of Planning and Gathering Evidence

Planning

Seema reported her planning by explaining how she started the activity. She said:

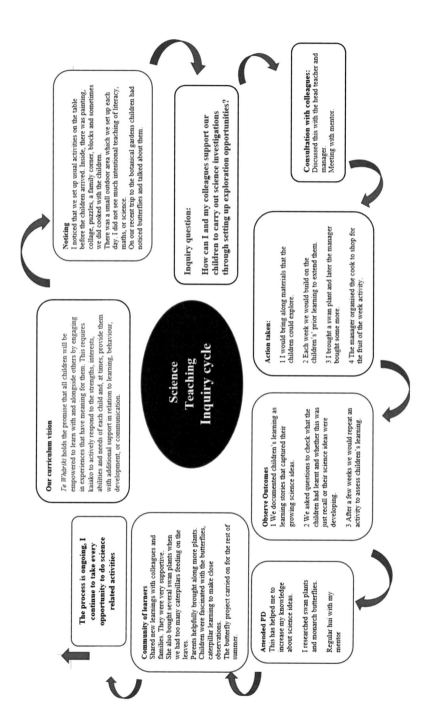

Noticing

I noticed that we set up usual activities on the table before the children arrived. Inside, there was painting, collage, puzzles, a family corner, blocks and sometimes we did cooked with the children.

There was a small outdoor area which we set up each day. I did not see much intentional teaching of literacy, maths, or science.

On our recent trip to the botanical gardens children had noticed butterflies and talked about them.

Our curriculum vision

Te Whāriki holds the promise that all children will be empowered to learn with and alongside others by engaging in experiences that have meaning for them. This requires kaiako to actively respond to the strengths, interests, abilities and needs of each child and, at times, provide them with additional support in relation to learning, behaviour, development, or communication.

Science Teaching Inquiry cycle

Inquiry question:

How can I and my colleagues support our children to carry out science investigations through setting up exploration opportunities?

Consultation with colleagues:

Discussed this with the head teacher and manager.

Meeting with mentor.

Action taken:

1 I would bring along materials that the children could explore.

2 Each week we would build on the children's' prior learning to extend them.

3 I brought a swan plant and later the manager bought some more.

4 The manager organised the cook to shop for the fruit of the week activity.

Observe Outcomes

1 We documented children's learning as learning stories that captured their growing science ideas.

2 We asked questions to check what the children had learnt and whether this was just recall or their science ideas were developing.

3 After a few weeks we would repeat an activity to assess children's learning.

The process is ongoing, I continue to take every opportunity to do science related activities

Community of learners

Shared new learnings with colleagues and families. They were very supportive.

She also bought several swan plants when we had too many caterpillars feeding on the leaves.

Parents helpfully brought along more plants. Children were fascinated with the butterflies, caterpillar learning to make close observations.

The butterfly project carried on for the rest of summer.

Attended PD

This has helped me to increase my knowledge about science ideas.

I researched swan plants and monarch butterflies.

Regular hui with my mentor

Fig. 3.2 Science teaching inquiry cycle

Our trip to the Botanical Garden was very helpful as the children saw monarch butterflies. We talked about their colour and when the children got close, they flew away. Children were so interested so we decided butterfly life cycle could be one science activity we could try. In the weekend I bought a swan plant (Teacher planning notes).

She also wanted to do activities that involved water play as it was spring and, on some days, children could play outside using the water trough. A discussion about the possible science exploration ideas took place at a mentor meeting as she had already begun to research and gather information about the monarch butterfly and how to grow butterfly's preferred swan plants.

Summary of butterfly investigation activities followed by Seema

1. We sowed swan plant seeds so the children could see the plants growing.
2. We looked at our seeds every Monday to see how they were growing.
3. Children learnt to use a magnifying glass to take a closer look at leaves and flowers.
4. Children noticed a monarch butterfly come to our big swan plant. After a few days they explored using the magnifying glasses looking for eggs. And we found and counted them.
5. Children loved reading the Very Hungry Caterpillar book. And we waited for the hungry caterpillars to come out of the eggs. And they did.
6. The caterpillars nearly ate up all our leaves so the manager took us on a trip to get some more plants from the garden centre. As soon as we arrived there Sam (Age 4) pointed to a plant and said, "Swan plant"! He could recognise it in amongst the other plants.
7. We had a summer of observing, counting, looking at caterpillars, and cocoons forming. Children were able to see a butterfly emerge from a cocoon and slowly unfold its wings.
8. Children's interest continued. One day, Nellie walked in with a dead butterfly she found on the way to the centre. We got our magnifying lenses out and made close observations. Children counted it had two pairs of wings, two antennae and six legs.
9. We watered our little swan plants and they continued to grow.
10. It was a great investigation to repeat next year (see Fig. 3.3).

Teacher reflection

As part of her development, Seema began writing her reflection. This helped her to think about children's learning, but also what she would need for the next day. The following is an example of the kinds of things she wrote in her notebook.

Our children know and love the hungry caterpillar story and we assume they know the life cycle of the butterfly. That is a story and their experience in this was with real, eggs, and how small the eggs are. How big the caterpillars grow and how many leaves they eat. We had looked at the picture of a caterpillar and counted the many legs they had. I had not thought about how in a picture children can only count the legs on one side. On our real caterpillars they counted all the legs. They got good at using the magnifying glasses and if they were not there, children asked us to get them. Little Sam who is two years old and not talking much

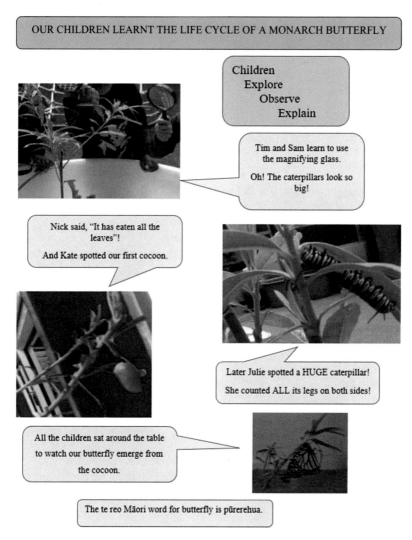

Fig. 3.3 The butterfly inquiry

yet spotted a large caterpillar, he made loud noises and pointed at the moving caterpillar excitedly (Year 7).

Other points she wrote included:

- Children come in and go straight to the swan plant and comment.
- Children and parents brought branches of swan plants from home to feed caterpillars.
- The manager got us more plants; other staff also used these to talk with the children.

- Caterpillars, cocoons, and butterflies became the subjects of drawing, painting, and collage. Jemma made a caterpillar from playdough and Sammy arranged the magnets into butterfly shape.
- Vocabulary development.
- Caring and being gentle with the caterpillars.

The mentor noted:

> Seema said that the children were keen to have butterflies back, so she had planted some seeds with them. "If we have the swan plants the butterflies will come again" (Mentor meeting notes, Year 7).

In sum, engagement, enthusiasm, and an understanding that caterpillars need the swan plants for a home and for food became apparent in the learning stories written during this time. And the butterflies came again.

As Seema gained confidence, she needed less and less help from the mentor. From the analysis of the learning stories, her interviews at the start and at the end of both years, her written reflections, and mentor notes, her development can be seen in Fig. 3.4.

3.7 Analysis of the Mentoring Data

Mentoring is designed to help develop mentee's expertise with the goal of induction of the mentee into the culture of the professions, in this case teaching. The mentor taking on the role of the coach may support the development of one or more job related skills or capabilities (Hopkins-Thompson, 2000). According to Bodoczky et al. (1999) mentors can adopt different roles, for example, as an educator, a model to inspire and demonstrate, an acculturate, or to induct the mentee into the culture of the profession. Mentor may also provide emotional support and be someone who listens when things are going well or not so well. Templeton et al. (2021) contend:

> Essentially, the mentor mentee relationship requires more than a combined willingness to succeed. Mutual responsibility and respect are foundational pillars, but even more crucial is having a growth mindset and a learning attitude. Mentors who are empowering are also great active listeners, are skilled at powerful questioning, and possess the ability to use self-reflection as a tool for resolute feedback that ultimately builds capacity in others (p. 349).

The purpose of mentoring in this research was to support a non-specialist ECE teacher to gain pedagogic competence. Dewi (2021) has found this approach to be successful with some teachers.

The mentor's diaries (63 hand-written pages) and teacher interviews (20, two each year) were analysed inductively, looking for commonalities within these from year to year. The following patterns emerged:

1. By year three Seema was no longer reluctant to try out new ideas suggested by the mentor. Mentor guidance was needed to try out new science ideas.

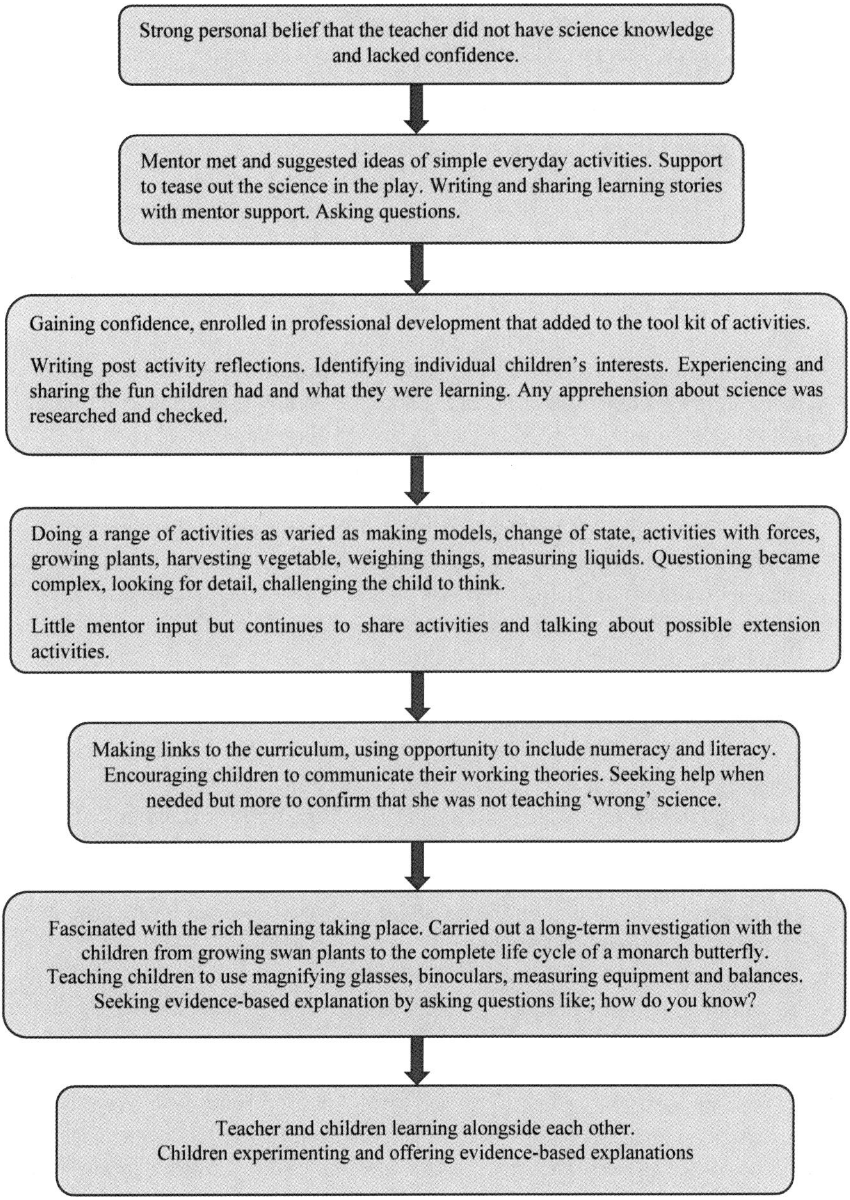

Strong personal belief that the teacher did not have science knowledge and lacked confidence.

Mentor met and suggested ideas of simple everyday activities. Support to tease out the science in the play. Writing and sharing learning stories with mentor support. Asking questions.

Gaining confidence, enrolled in professional development that added to the tool kit of activities.

Writing post activity reflections. Identifying individual children's interests. Experiencing and sharing the fun children had and what they were learning. Any apprehension about science was researched and checked.

Doing a range of activities as varied as making models, change of state, activities with forces, growing plants, harvesting vegetable, weighing things, measuring liquids. Questioning became complex, looking for detail, challenging the child to think.

Little mentor input but continues to share activities and talking about possible extension activities.

Making links to the curriculum, using opportunity to include numeracy and literacy. Encouraging children to communicate their working theories. Seeking help when needed but more to confirm that she was not teaching 'wrong' science.

Fascinated with the rich learning taking place. Carried out a long-term investigation with the children from growing swan plants to the complete life cycle of a monarch butterfly. Teaching children to use magnifying glasses, binoculars, measuring equipment and balances. Seeking evidence-based explanation by asking questions like; how do you know?

Teacher and children learning alongside each other. Children experimenting and offering evidence-based explanations

Fig. 3.4 Teacher development from a novice to a confident teacher of science in ECE

2. In year five, her children were using magnifying glasses to look for detail and she was very excited when a child asked for the magnifying glass to make things bigger to see!

3. Seema's confidence grew, and she began to believe that she was learning science and helping children to learn science. In year 7, the mentor noticed that Seema was being able to think, plan, and carry out many activities. She was using learning conversations with the children for assessment and being able to plan to extend them.

4. Although the mentor had guided her to use natural world examples and everyday objects, by year 8, she was asking what else would be useful.

5. The introduction of a revised curriculum (Ministry of Education, 2017) impacted on her confidence. The shift from following the child's interest to being able to do intentional teaching took some time for her to come to grips with the changes.

6. In year 9, she had figured out with mentor guidance how to use the curriculum to help children develop their own theories and to refine them. This was the time when help was needed to introduce ideas like forces and magnetic properties.

7. In year 10, she had become very confident to come up with her own ideas, which she checked with the mentor but was able to implement, extend, and assess all by herself. Now she had a diverse portfolio of learning activities.

8. While she worked in three ECE centres during the research. It was also noted that from year 5 onwards she was able to competently integrate whatever the focus of learning was for her centre at the time, alongside science learning. For example, with whānau support in providing information, her children learnt to say their mihi (Mihi is the Māori way to introduce oneself). Similarly, in in year 6, she was responsible for the 4+ age children for their transition to school. There was a focus on oral and written language and children learnt to write the alphabet and their name before starting school. In year 9, in a different ECE centre there was a focus on bilingualism, and she was able to integrate te reo Māori in her intentional teaching.

9. It is significant that the constant introduction on new terminology still shakes her confidence. For example, she likes to talk through new documents that are put on the Ministry of Education website. For example, the recent Te Kōrerorero Talking together resource. It uses strategies that are called 'serve and return' for asking questions and children answering them to continue a conversation.

10. A related issue has been how the three ECE centre's she has worked in during the period of the research understand and implement 'Teacher inquiry'. This appears to come from the understanding that the head teacher has of what this ought to be. Each centre has followed their own template with one centre that was involved in research creating a template that was 10 pages long! The adjustment to the next centre's A3 size template became a challenge for her! As recorded in the following notes.

It is noteworthy that the three centres that Seema worked at all had different approaches to teacher inquiry. One where the inquiry was collectively done by all staff on a common topic. In the second centre each teacher conducted their individual inquiry and followed a ten-page long template. In the third centre a template was provided with guiding questions to be followed (Mentor notes based on teacher interview, Year 9).

3.7.1 Mentor Reflection Analysis

The mentor had previous experience of mentoring new teachers as a head of science department, which required a short-term of mentoring. She had also mentored student teachers she was educating for 20 years that overlapped this research period. She reflected on change in her mentoring practice over the years and wrote in her diary.

> I think mentoring Seema has helped me to become a better mentor. I now know when I just need to be a listener for her to talk aloud her ideas and answer her own questions. This I have noticed has also helped me to mentor my students beyond their teacher training. For example, they can contact me when they need support, access new resources or just get affirmation for what they are thinking to try out (Year 5 diary entry).

> .... An added advantage has been the change in what I thought was mentoring. I have read and attended PD sessions to update my own knowledge. Mentoring does not have to be face to face contact, although that is great. Supporting my former students through email or phone call has helped me to set up a network of science teachers. I know their interests and can send them pdfs of my own publications as well as other relevant reading that they no longer have access to once they have left the university. This keeps them in touch with recent research (Year 8 diary entry).

The mentor asked a colleague involved in ECE teacher education to check the analysis of the mentor diary notes to critique the analysis. This has provided confidence in the analysis process.

Going over and over the notes of the length of the study, several questions have arisen for the mentor.

1. How effective are the current practices of teacher professional development?
2. Why is there a compulsive need to add new terminology which is not unpacked?
3. What teacher support is provided when transitioning into a new curriculum which has significant changes?
4. How effective is putting information on the Ministry of Education website for teachers to read it, and come to a shared understanding of creating a local curriculum?
5. In what ways would timely mentor support be better than collective professional development sessions?

These thoughts are unpacked in the mentoring theme in the discussion chapter.

3.7.2 Analysis of Teacher Interview Data

The interviews followed a semi structured interview schedule and were conducted informally, allowing Seema to talk and ask questions. The first interview each year was focussed on what Seema was thinking on doing during the year, clarification of the focus set at her centre, and considering what she was going to do her inquiry on in that year. The interview times varied and took approximately the time noted in Table 3.1.

Generally, the time taken decreased as the years progressed with an anomaly in year 7, when the new curriculum was implemented. Year 7 also required two extra one-hour meetings for her to start using the curriculum and to discuss the learning stories that she had written to the requirements of the new curriculum. The second interview was a review and reflection session which was useful to draw her attention to the confidence she had gained, what her children had enjoyed the most, what she found interesting and any challenges she faced. The following is a summary of the trends that emerged.

- Confidence grew consistently throughout the 10 years.
- The range of science explorations widened. For example, investigating floating and sinking (Year 2), ice activities (Year 3), using magnifying lenses (Year 5), taking children to botanical gardens to play and explore (Year 5), Trip to Te Papa Museum to learn about bugs and Matariki and exploring magnets (Year 6), Exploring snails, insects, spiders (Year 7).
- Learning to use te reo Māori greetings and for instructions (Year 4 onwards). Including an inquiry on teaching and learning the Māori language in her centre.
- Reading science related books to the children, for example Who sank the boat? A story related to floating and sinking (Year 5).
- The learning stories increasingly focussed on encouraging children to talk and including numeracy, vocabulary development, extending learning, and assessment of learning (Year 6).
- The complexity of the activities increased. For example, if earlier on children sorted things that would float or sink, later they were encouraged to consider how they could make something that sinks, to float. Similarly, children exploring using a telescope and later being asked to compare how it was different to using magnifying lenses.
- One noticeable feature was how any changes in the centre policy, or the curriculum and its implementation was unsettling for her and shook her confidence. She needed more guidance when the curriculum changed and each time a new format was provided for teacher inquiry.
- Her passion for science teaching and learning has grown exponentially.
- Seema is increasingly using the novel exploration opportunities to help settle children when they arrive and do not want their parents to leave. Snails and flowers were specific examples she used in several interviews.

Table 3.1 Approximate time taken for the interviews in each year

Time/Year	1	2	3	4	5	6	7	8	9	10
First interview time	1:40	1:42	1:35	1:30	1:20	1:30	2:00	1:20	1:20	1:00
Second interview time	1:30	1:40	1:25	1:30	1:20	1:25	1:40	1:30	1:20	1:10

3.8 Summary

This chapter has presented the analysis of data with respect to teacher development. This included teacher interviews, reflections, learning stories, as well as the mentor data from notes and diary entries. As with all learning journeys, this one began with the teacher taking the initiative to learn because she was encouraged to start with something which was familiar. The reaction of children to the simple first activity of exploring fruits and knowing that she knew what she wanted to teach gave her the confidence in her own *content knowledge*. The development from there onwards was her personal motivation and a real desire to challenge herself and her children to think. It is like the first step that a child takes. Parents look forward to it and it is a big deal. If one of them misses seeing that first step, they feel they have missed out. Why is that first step so important? We believe that with it comes the promise of where the next step will take the child. This has been perhaps the case here too.

References

Bodoczky, C., Malderez, A., Gairns, R., & Williams, M. (1999). *Mentor training: A resource book for trainer-trainers.* Cambridge University Press.

Cherrington, S., & Thornton, K. (2013). Continuing professional development in early childhood education in New Zealand. *Early Years, 33*(2), 119–132.

Dewi, I. (2021). A mentoring-coaching to improve teacher pedagogic competence: Action research. *Journal of Education, Teaching and Learning, 6*(1), 1–6.

Halbert, J., & Kaser, L. (2013). *Spirals of inquiry.* BC Principals and Vice Principals Association.

Handscomb, G., & MacBeath, J. (2006). Professional development through teacher enquiry. *Journal Issue, 1.*

Hopkins-Thompson, P. A. (2000). Colleagues helping colleagues: Mentoring and coaching. *NASSP Bulletin, 84*(617), 29–36.

Lammert, C. (2020). Becoming inquirers: A review of research on inquiry methods in literacy preservice teacher preparation. *Literacy Research and Instruction, 59*(3), 217. https://doi.org/10.1080/19388071.2020.1730529

Ministry of Education. (1996). *Te Whāriki: He Whāriki Mātauranga mō ngā mokopuna.* Learning Media.

Ministry of Education. (2007). *He New Zealand curriculum.* Learning Media.

Ministry of Education. (2017). *Te whāriki. He whāriki mātauranga mō ngā mokopuna o Aotearoa: Early childhood curriculum.* Retrieved from https://www.education.govt.nz/assets/Documents/Early-Childhood/ELS-Te-Whariki-Early-Childhood-Curriculum-ENG-Web.pdf

Timperley, H., Kaser, L., & Halbert, J. (2014). *A framework for transforming learning in schools: Innovation and the spiral of inquiry* (Vol. 234). Melbourne: Centre for Strategic Education.

Templeton, N. R., Jeong, S., & Pugliese, E. (2021). Editorial overview: Mentoring for targeted growth in professional practice. *Mentoring and Tutoring: Partnership in Learning, 29*(4), 349–352.

Chapter 4
Research Evidence of Children Playing and Learning to Investigate

Abstract Children come to understand the physical and natural world around them when given the opportunity to explore and play, sometimes by themselves and at other times with their friends. Children can come to understand and make sense of their world in different ways. In our bicultural country children are to be encouraged to understand the world both from a Māori and a science perspective. In this chapter, we share the rich learning opportunities that the teacher provided for the children to experience the physical and natural environment around them and how they made sense of the physical and natural world from a scientific perspective.

Keywords Analysis of learning stories · Mentor data analysis · Children learning to investigate

As the teacher gained knowledge and experience, her planning became increasingly purposeful. She was herself inquisitive and enjoyed trying out new ways of engaging children in play. Here we share different approaches to investigating the environment that the students experienced. Evidence of children learning how to investigate by sorting and classifying, exploring, looking for patterns, understanding fair ways of investigating, making models, and even developing systems is presented. We also share teacher planning and reflection that indicated teacher's thinking about future learning activities.

4.1 Introduction

International research shows that children's scientific experiences are limited and dependent on the teacher's knowledge, attitudes, and beliefs (Freeman, 2021; Pendergast et al., 2017).

In Australia, Fleer et al. (2014) reported that much of the science and other teaching practice in ECE in the 1980s was informed by a developmentally appropriate practice (DAP) which was informed by the works of Piaget and Bredekamp. Consequently, much teacher preparation and professional development was embedded in this approach. Later MacNaughton (1995) questioned whose worldview was being privileged by this approach and brought into consideration the cultural lens through which children may develop their understanding of the world. Resulting in rethinking of the stage-based model in Australia and in the United States (Clyde 1995; Fleer & Kennedy, 2000). In contrast to the practice of following the child's interest, a practice common at the time Klahr and Nigam (2004) asserted that children learned more from direct instruction than from discovery learning and were able to make richer scientific judgments. Which does not find agreement with Hirsh-Pasek et al. (2009) who proposed that discovery-based, active learning is a powerful pedagogical approach. Following further research Hirsh-Pasek et al. (2015) claimed that learning was enhanced when adults provide learning environments that encourages fun-led exploration and discovery. Such an environment was repeatedly provided by the teacher in this case.

Fleer (2014) suggested that authentic scientific learning takes place when scientific problems are introduced while teachers engage in role-playing scientific concepts with children. We think, for example, if the teacher models that the container she is using to fill the bucket needs 20 cups of water, it raises the possibility of children counting with her. But more than this if asked how we can fill it up faster, the children are encouraged to think of alternatives such as using a bigger jug instead of the cup. Fleer (2014) recommended pedagogical approaches that combine play and scientific narrative in play-based settings. She added that these unique pedagogical characteristics promote scientific narratives in play-based settings.

Fleer and Prambling (2015) suggested that the research at the time did not differentiate the pedagogical practices which link with "imagination in science and imagination in play (p. 1259)". Later Fleer (2019) reported that when children are engaged in play, creating scientific narrative alongside 'wonderings' were important factors of science learning in play-based settings. Imagine a child being asked to move a box from one place to another. Having a play trolley available along with the question, how can we move this box easily, may well lead to trial and error. Children may try pushing the box or pulling it along and maybe let us put it in the trolley. Children are more likely to make the connection between observable and non-observable contexts. Such as the magnetic play activity presented later in this chapter.

In 1995 New Zealand mandated Te Whāriki (Ministry of Education, 1996) our first bicultural curriculum that had two parallel documents, one in Te Reo Māori, used in Kaupapa Māori Based Kohanga Reo and the English version for other ECE Centres. Te Whāriki emphasised the "critical role of socially and culturally mediated learning and of reciprocal and responsive relationships for children with people, places, and things" (p. 9). Importantly it brought forth the idea children's learning is a collaborative process. Further, that children learn "through guided participation and observation of others, as well as through individual exploration and reflection" (p. 9).

Te Whāriki (Ministry of Education, 1996) our first bicultural curriculum had two parallel documents, one in Te Reo Māori, used in Kaupapa Māori based Kōhanga Reo (Education based settings where Te Reo Māori language is the primary language mode of learning and teaching) and the English version for other ECE Centres. Te Whāriki emphasised the "critical role of socially and culturally mediated learning and of reciprocal and responsive relationships for children with people, places, and things" (p. 9). Importantly it brought forth the idea children's learning is a collaborative process. Further, that children learn "through guided participation and observation of others, as well as through individual exploration and reflection" (p. 9). Freeman (2021) recommends:

> Teachers can choose to notice, recognise, and respond purposefully to children's scientific interests or to happenstance provocations that come into their centre and take those scientific enquiries to a deeper level. Using planned provocations to extend children's interests or bring in new investigations can facilitate further explorations and wonderings (p. 34).

A provocation could be something a child can engage with, look at, handle, or draw that supports children's exploration and wonderings.

It is noteworthy that activities are set out, children are allowed to play and explore freely. Seema uses the engagement for encouraging children to think and come up with their working theories from their observations. As intended by the curriculum, language development, numeracy and oral communication are woven in. Language and numeracy development and oral communication are required in the revised Te Whāriki.

4.2 Presentation of Learning Stories and Teacher-Learner Learning Conversations

Our curriculum, *Te Whāriki*, has five strands of Wellbeing, Belonging, Contribution, Communication, and Exploration. Science investigation provides opportunities for exploration, where children participate and contribute ideas and develop the skill of communicating in English and Te Reo Māori. Here we present the analysis of a teacher evaluation of a teaching and learning sequence and related conversation. It illustrates the diversity of learning that takes place during play when the teacher deliberately plans activities that attract, engage, and create possibilities to play and think.

The learning stories presented here have been selected from over 200 stories the teacher shared with the mentor. Of these, 50 stories provided examples of science learning experiences. A pragmatic approach was taken in selecting the narratives as learning stories and learning conversations with two purposes. First, providing evidence of children engaging in the different approaches to investigation. Secondly, these stories would also be helpful for other teachers who may want to replicate them in their own ECE environments.

The learning stories and learning conversations presented illustrate that when children are given novel materials to explore, they begin to learn scientific ways of investigating their world. The teacher learnt that providing rich experiences that attract children to engage and play was important. It was also critical to allow them to play freely and observe, then playing alongside them, asking question to encourage conversation.

4.2.1   Children's Experiences of Different Approaches to Investigation

Six different types of school investigations are described internationally (Watson et al., 1999; Windschitl et al., 2008). The New Zealand Curriculum (Ministry of Education, 2007) calls them approaches to investigation and the expectation is that the students will learn these approaches to investigation during their schooling. We describe each and what it looked like in our study ECE centre.

1. classifying and identifying,
2. pattern seeking,
3. exploring,
4. investigating models,
5. fair testing,
6. making things or developing systems.

Each approach to investigation is described, followed by one or more examples of learning stories or learning conversations. Next, any reflections noted by the teacher are presented. Finally, a comment is made on the learning that has taken place.

1.   *Classifying and identifying*

Being able to identify different things is a skill needed to manage everyday life. Simple activities like sorting building block by colour or size or sorting the washing at home for each family member are everyday examples of classifying. When children go to the supermarket, they learn that things are grouped together. Fresh fruit and vegetables are in one place, frozen things in another, for example. They soon learn to identify the isles where different things are. If their family mostly visits one supermarket, they can help their parents locate things even before they have learnt to read. Most identification at this stage is by the images they have seen whether it is the breakfast cereal or a block of butter. Life would be a lot harder if the supermarket put everything in the middle in a pile and asked them to find what they needed. Sorting and classifying teaches two important skills, to look for things that have similar properties and to organise them.

The following learning story shows how children are engaged in water play and sorting things that float or sink in water.

Will it float or sink?

Learning story

> Children love water play so a bowl of water and an assortment of objects were put out for the children to play. Children put things in water. They loved pouring water from one jug into a bowl. As children were busy playing, they were asked which things they thought will float and which ones will sink? We started the activity with children predicting, explaining, observing, and expanding. At the end of the activity, children looked at things that float and concluded they were all made of plastic and were light. Metal and rock and other heavy things sank. Mira found that the hard rock sank and volcanic rock which is pumice floats. She also found pumice is lighter than hard rock. We talked about how pumice forms. Sebastian found a stone and each time he put it in, it sank. When asked how he could make the stone float Seb became excited to show how it worked. He put the stone on a plastic lid, but he was in a rush to do this, and it sank. But he did not give up, tried again, and put the stone on the lid, and it floated! Seb continued exploring. He put a metal bowl in the water, and it floated. He put other things that were sinking into the bowl and made them float and had fun making the sinkers float. When asked, why things did not sink, he said, 'It is like a boat' and moved on to play outside.

This learning story shows that children are engaged in hands-on exploration. The teacher provided the opportunity for children to compare, which requires higher-order thinking. Seb had clearly applied his prior knowledge of boats floating on water to change the sinkers into floaters. This activity gave the children opportunity to play, explore and sort things into floater and sinkers.

Sorting shapes

> Today I had put out different shapes on the table. After cutting up the shapes Joe decided to paste all the triangles of different colours on one page. Noah cut out all the circles and carefully arranged them on the table. Other children were just happy to talk about the colours. I encouraged them to use both English and Te Reo Māori names for the colours.

Similarly, when allowed to play with a pile of leaves, children were able to group them into big and small. Green and not green and pointed and not pointed. Other examples of sorting activities used were children sorting fruit that were long or round, juicy, or crunchy and ones that needed to be peeled before eating. Playing with blocks and toy cars and trucks can be extended for getting the children to sort them according to size.

2. *Pattern seeking*

Looking for patterns helps children to make sense of things around them. The idea of using a particular property to organise things in patterns, teaches children to compare things. Whether it is differentiating from brightest to darkest colour, tallest to smallest doll, or smallest to the largest ball.

The leaf sorting activity was extended by asking children to arrange them from the longest to the shortest (see Fig. 4.1). Similarly, while playing with dolls, they are asked to line them up from the smallest to the tallest. Being able to look for patterns is a useful skill both in science and in mathematics.

Interestingly, the children have organised them according to size whether it was a single leaf or a branch with many.

Fig. 4.1 Leaves arranged from tallest to smallest

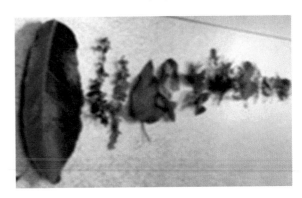

3. *Exploring*

Very little children start exploring things by holding them or putting them in their mouth. A little child manages to get their big toe in their mouth! Similarly, once crawling they get into everything! They need to explore to make sense of everything around them. As is evident, Seema gave the children many opportunities to explore. The activities included: states of matter, melting and freezing (ice and water activities); forces, exploring magnets pushing and pulling, for example.

Exploring feathers

Learning story

> Our children had explored bird nests on the previous day. As we had talked about birds, I brought feathers for them to explore. Children looked at the long and colourful feathers and felt how light they were. They had a close look at the tiny feathers of a finch and fantail. These felt very soft. They decided that the gull's feather was hard and strong and so was the wing feather of the peacock. Children played with the feathers and tickled their friends and themselves with the feathers. Our tamariki (children) were gentle when they held the feathers. They also used magnifying glasses to have a closer look.

Teacher reflection

I think it would be good to bring the three nests for them to explore. Then we can talk about bird's homes and how the birds make their nests.

Such sensory exploration allows children to play with natural materials that they may not have contact with routinely and helps them come up with their own theories about why birds have feathers and how they use them. As children talk, words like light, long, colourful are introduced. By this they are developing their language, using new words which supports their literacy and communication skills. It also provides opportunity for talk about why birds have feathers, for example, for warmth, flight, and to look nice!

Exploration is play at its best! It allows children to be creative and as they talk with each other, and the teacher make sense of the everyday things in life as shown in the following exploration activity and beautiful creative arrangement of flowers (see Fig. 4.2).

Fig. 4.2 Creative arrangement of petals by a child

Playing with magnets an attractive and extended exploration

Children are familiar with magnets and enjoy using fridge magnets and magnetic letters and numbers. The following story is based on one such day of exploration.

Learning story

Today's exploration was with magnets and assorted things made of different materials. There were things made of plastic, steel, iron, a paperclip, a piece of shell, a candle as well as a horse-shoe magnet. This attracted the children, and they came along to look and play. Richie, Rina, Sam, Amy, and Hemi came and sat around the table and were looking at the things. Children are familiar with magnets because they play with magnetic shapes, and magnetic letters most likely on the fridge at home. They were excited that today they had so many things to play with! Everybody had magnets and they were exploring, touching the things with the magnets. They were asked to try and work out which things will stick to the magnet and which ones do not. When asked why some things were not attracted by the magnets, the answer was because they were not metal. Cam was excited to show us that a bottle top lid was moving on the table when he moved the magnet under the table. Then I put some iron dust in a sealed container and steel wool in a bag on the table. The children had two bar magnets to try and move these. They loved playing and seeing the iron sand and steel wool move above the table.

Children were able to put two magnets together and Hemi was intrigued that it made the magnet stronger. When asked to push the two ends of the magnet together they tried hard, but the magnets just would not stick together! That was a lot of fun. Later we talked and I explained to them that magnets have two ends that are called north and south pole. When we bring the north pole of one magnet close to the south pole of the other, they are attracted and easily join. But when we bring north and south poles together, they push each other away. Through trial-and-error children found out that some things stick to magnets and others do not. Their working theory was that things made of metal are attracted to the magnets. To help them refine their theory I gave them my chain and told them it was made of gold which is a metal. When they tried, the gold chain was not attracted to the magnet.

Teacher reflection

To help children understand and refine their theory that only things made of iron are attracted to the magnet. I want them to explore and think deeply about what they have learnt. Next time I will give them other metals for example silver, copper, and aluminium to try out.

Children love playing with magnetic shapes. They build lots of patterns and structures. They already know that magnetic shapes stick to each other, and I have noticed that children find that some shapes repel while they try to connect them together. They learnt the reason for this from the above learning story about the two magnetic poles.

4. *Investigating models*

Models are not just products of science but also tools and processes of science (Meng-Fei & Jang-Long, 2015). Models have been used for explaining, predicting, making a scientific phenomenon visible. For example, a model can be used to look at human body parts. In ECE the simplest form of models are railway tracks and carriages, emulating how trains move. They can also be used as thinking and communicating devices for testing science ideas. Often, children-built models that illustrate an abstract idea in their mind as a concrete version of it. The process of scientific modelling is an iterative process of construction, evaluation, and revision or replacement of models (Meng-Fei & Jang-Long, 2015). Those working in ECE would have often seen children building a model of a racing track, then when they want to show the cars going around the bend or over a bridge, new bits are added. In a way, at their early stage of learning the child is in control of its development. As they play and think, sometimes have input from the friends, they know they can change, remake, or even destroy them if the model is not doing what they intended. We consider models to help children understand the non-linear and iterative nature of science.

Playing with models and exploring properties of magnets

As children had loved playing with magnets, following their interest, the teacher brought some magnetic toys for them to explore the magnetic properties. The box contained, toy cars with built in magnets, couple of toy people also with magnets, some magnetic discs, sticks and a sealed container that has some iron sand. When she set up the toys, children came along and sat on chairs. Mike and Anuj chose the cars, and she distributed the people and the sealed magnetic sands container to the other children. When they finished playing with their toys, they were asked to swap their toys with others. Mala was happy exploring the magnetic people.

> Anuj tried to stick two yellow sides of the circle to the yellow sides of the magnet bar, but he found purple and yellow sides stick together. Mala and Sue found that two yellow sides of the disk pushed each other away (repelled) but different colours stuck together. I heard

Hasan notice that the disc magnets are like donuts. Cam and Moi noticed other children playing and patiently waited for their turn. Cam moved the car backward and forward not touching the car. While exploring the magnetic toys, children had a lot of fun. They worked out that when two magnets were brought close to each other, they either pulled together or pushed each other away. They also worked out that they could move the toy cars backwards and forwards without touching them.

Teacher reflection

This activity gave children the opportunity to play cooperatively with their friends, sharing and taking turns is a valuable social behaviour that our children are becoming very good at. This play helped children to try out different ways to move the toys. They considered and did problem solving and suggested reasons for why the toys were either pulling together or pushing apart. These are magnetic properties and shows how magnetic force works.

Making models

Learning story

Our children have been used to making models of volcanoes in the sandpit. They have also made things using clay and playdough. To celebrate the Indian festival of lights (Diwali) our children made little earthen lamps from clay. The cook baked these for us in the kitchen oven. Then we put oil in the clay lamps and children used some cotton wool to make wicks. We put oil and lit our lamps for our celebration.

Teacher reflection

Our children also like building with blocks. Having conversations with them about their constructions is insightful. They have stories to tell about their models. Sometimes, these are castles, and at other times railways with trains on tracts. Rewi often arrives and goes to the blocks. I need to talk with him about his model making.

Creative Rewi built a robot and explained how it works

A learning conversation

This learning conversation shows Rewi's creativity and how the teacher finds out what his model is about. It also provided the opportunity for a learning conversation and encouraging him to draw on prior knowledge about birds.

When Rewi arrives at the centre, he heads to the construction area. He may choose to play with magnetic shapes, mobile, lego, plastic or wooden blocks. Recently, he had favoured playing with mobilo (building blocks). Today he built a structure that he explained was a robot. This provided an opportunity to have a conversation. He pointed to his model and explained that it could fly. We talked.

Teacher: "Does it use its hands to fly?" (Pointing to the structures that looked like hands?)

Rewi: No and pointed to the hands which were in the front. He explained that those two things help robot to fly but he did not know what they were.

To extend our conversation...

Teacher: Which animal flies? What have you seen flying around outside and in the garden?

Rewi: Birds

Teacher: What special body parts birds have that help them to fly?

Rewi: Wings. He explained that his robot had two wings and pointed to these.

He was patient in trying to explain how his robot could use its wings to fly. Clearly, the teacher needed explanation!

Teacher: Rewi what can your robot do?

Rewi: My robot can fly over the mountain and jump up too!

This story shows that Rewi is creative and is developing his oral communication skills. The strategy used to extend the conversation was offering a similar example to help him remember that birds fly and then he was able to add that the Robot used wings.

After this conversation, Rewi confidently explained his model and how it works to his younger brother.

Teacher reflection

Rewi was interested in insects for a long time. His friends supported his interest by calling him as soon as they saw an ant, a bug, or a butterfly or even a spider. Although he is still interested in the little creatures his new interest is to play with blocks. Learning conversations with him have helped to improve his oral communication skills.

5. *Fair testing*

A common cry when asked to do something that a child does not want to do is "this is not fair"! However, the concept of fairness is not that easily learnt. Taking a moment to recognize and explain to the child when they are not being fair is a useful practice. Helping them to notice fairness in their own behaviours will help their understanding of the concept. Fairness may not be taught as easily but with consistency, children will recognize fairness and act fairly towards others. In ECE, teaching turn-taking, modelling fairness, drawing attention to fair and unfair behaviours, and praising fair behaviour are some ways of teaching fairness. Explaining to the children that it is only fair for everyone to get to do something, or it is fair to wait for their turn sets the foundation of what fair testing is based on.

Science investigations involve posing a question, testing predictions, collecting, and interpreting evidence, drawing conclusions, and communicating the findings. When scientists conduct a fair test investigation it is to answer a question or to test a prediction.

Good examples in everyday life can be asking children to compare two types of apples and to mention that comparing apples with bananas would not be fair because they are different fruits.

The idea is to make comparison by keeping all factors but *one* the same. Having two plants of the same size, keeping them in the same place, but getting the child to

Fig. 4.3 A visual example of fair testing

water one and not the other may help the child understand that plants need water. However, explaining why, we are using two plants the same and having them in the same place is to begin the thinking that water was the only thing different so is likely the reason for the plant that was not watered dying.

This can be seen in the following picture where two flowers of similar size were placed by the children in jars. Jar A did not have any water whereas children put water in Jar B. Each day they observed the two flowers. At the end of the week, they could see that the flower in Jar A was drying out and that in Jar B still looked healthy (see Fig. 4.3).

In a fair test, all other variables must be controlled. These variables are known as the controls (or controlled variables).

6. *Making things and developing a system*

Although children automatically try to make sense of our world, their common-sense explanations are often loosely collected. The disconnected nature of everyday thought is different to scientific sense making that relies on coherent and systematic explanations (Penner, 2000). One way that we can encourage the notion of developing systems is by encouraging them in modelling phenomena. Making models, testing, and evaluating their model of phenomena is how scientists formulate explanations (Penner, 2000). For children to explain phenomena they too need to have the freedom to explore, make models, sometimes abandon them, think more, and have another go. Working together brings in new ideas and can support the continuation of the developing system. Easy access to materials makes this task easier. For these reasons we notice more systems thinking and model making in sandpits where there is access to lots of resources or when children have access to lots of different building blocks. The following learning story shows such an activity in the sandpit.

Children make a volcano in the sandpit and talk about it

Mona, Livisi, and Sara were busy making a volcano in the sandpit. Mona and Sara were digging, and Livisi was pouring water into the crater of the volcano. Sara explained that water was coming out of the volcano. To give them an example of volcano which releases hot water I talked to them about Mt. Ruapehu and how dangerously hot the water is. While we were talking children were continuing their work. Mona dug sand to make a river all around the volcano while Livisi continued to pour water in the crater of the volcano. Gradually they made a wide river around it. We talked about their volcano looking like an island. Using the example of Somes island in our Wellington Harbour made it easy for some of the children to visualise. Then I asked how we can cross the river to reach their volcano mountain. Mona said ''we can fly there''. During our conversation Mike brought a boat and made a bridge with a piece of wood. He also made a ramp to make it easier to go on the bridge. By this time Nelly and Tatum joined this group. They also contributed by digging and bringing water. Livisi explained to them that they were making an island and he explained that "An island is a land surrounded by water". Good on you Livisi for describing what an island is to your friends who had just joined us.

This story shows that children are developing the social skills of working together collaboratively, contributing their ideas in a shared project. They have learnt new words, for example, crater and island. Significantly, they are beginning to develop the idea of a connected system. The volcano erupts, the lava flows, it builds up around the mountain and now we have an island. There was opportunity to solve the problem of getting there. Mona suggested flying there as an option and later Mike brought a boat and built a bridge. The teacher question, how the river will be crossed made the child think, first a quick response, fly there. Followed by continuing to develop the system further by adding a boat and a bridge an indication of deeper thinking and creativity.

4.3 From Situational Interest to Personal Interest

Over the course of the research there were many examples of children engaging in activities that led to a more enduring personal interest for the children. As she gained experience, Seema became increasingly conscious of the need to provide learning opportunities for children that were perhaps not accessible at the centre. These included trips to the National Museum which was close to one ECE centre where she had worked and walks in the botanical gardens when working at another.

Here is a summary of a few experiences that led to personal interest for the children.

- When children did not have green spaces to play, she organised trips to garden centres. On one such trip, one four-year-old identified a swan plant (milk weed) because they had them at the centre for the monarch butterflies to visit and lay eggs.
- By taking insects that she found around the home and teaching children how to use a magnifying glass, children developed an interest in making close observations. Several children would ask for the magnifying glasses often to use them.
- Similarly, while working in a centre where children did not have expensive building blocks at home, she made sure they had access to these at the centre regularly. This encouraged one boy to arrive and head straight to the block area.

- Children grew sunflowers, beans, other vegetables in another centre. They reminded her that they should grow swan plants "otherwise the butterflies won't come".

We present one such experience that led to a deep personal interest in snails for one child.

4.3.1 Exploring Snails, Creating Situational Interest

The teacher found some snails in her garden as the children had been interested in learning about insects, this was an ideal opportunity to extend their learning to other small animals. This story also illustrates Fleer's (2021) idea of imagination in play and the notion of conceptual play worlds.

A group learning story

> Tama said that he liked snails. I brought two snails from my garden. I put them on the table and the children came to have a look. Tama was also keen to take a closer look. He and the other children used a magnifying glass to have a look at the snail's shell and its eyes. They liked looking at the snails and talking about its legs. Then they decided that snail did not have any legs. Great observation, as the children have been learning about insects and know that insects have six legs. When we pointed to the tentacles Jacob said that they were called tentacles and that the eyes were at the end of the tentacles. The children continued to explore for some time. They noticed that the snail leaves a trail behind. Eva said it has a shell. I explained that the shell protects the snail from danger. When they are afraid, they go into their shell. Snails come out at night so that the birds can't kill and eat them. Tama said that there was a mummy and baby snail. They all agreed that snail moves very slowly. It was interesting to see Martin acted like a snail. Following Martin's role play other children acted like snails and had a lot of fun.

> Kiri gently picked a snail and put it on her hand to take a closer look. She was not scared and did not react when the little animal started moving. I was so proud as she started exploring and enjoyed the tactile experience of having it move on her hand. We talked about the body parts of the snail.

Teacher reflection

This activity gave children the opportunity to explore and learn about snails. Some children contributed their previous learning which helped others to learn. And for those who had previous experience their learning was affirmed. We also used this opportunity for learning te reo Māori including learning the name ngata for snail. I think the children had a lot of fun and new words were introduced. It would be good to bring the snails again to assess what the children have retained and possibly talk about what it eats.

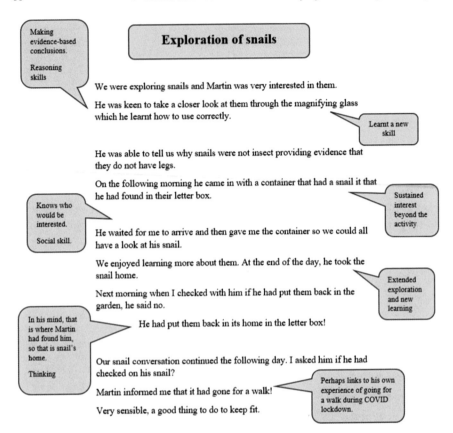

Fig. 4.4 Martin's snail exploration

4.3.2   Martin's Personal Interest

This story shows that by providing snails the teacher created situational interest which helped develop Martin's personal interest in snails (see Fig. 4.4).

4.3.3   Cameron's Personal Interest in Growing Sunflowers

In the last year of this research, children explored seeds of different kinds and planted sunflower seeds. They made observations each day and watered their pots. There was much excitement when the seeds germinated. Children continued to care for their seedlings and later planted them into pots and in the centre's garden. Some children wanted to take their plants home and look after them. Cameron took his plant home

Fig. 4.5 Cameron grew a
sunflower from seed

and cared for it and the following picture was sent by his mum when the plant
flowered (see Fig. 4.5).

Our curriculum encourages seeking whānau (family) support and this helps in
children's learning being reinforced at home.

4.4 Summary

The analysis of the learning stories and the evidence shared here is exciting as over
time the pedagogical approaches taken by the teacher led to the children experiencing
a 'library of experiences' and learnt from all different types of investigations ((Watson
et al., 1999), not just once but repeatedly.

This chapter has provided a comprehensive account of the playing and learning
opportunities provided by the teacher. Evidence presented shows that with thoughtful
and intentional planning and teaching very young children can experience and learn
the multifarious ways in which science investigations can be conducted. We are aware
that this selection has not presented the rich language, numeracy, and te reo Māori
development that took place alongside this science learning. This we have balanced
against the need to focus on the gap in the literature and indeed learning materials
available for children's science learning. The next chapter discusses the themes that
have emerged from the integration of all analysed data.

References

Clyde, L. A. (1995). Resources on the internet for management and professional development. *Scan: The Journal for Educators, 14*(4), 46–49.

Fleer, M. (2009). Understanding the dialectical relations between everyday concepts and scientific concepts within play-based programs. *Research in Science Education, 39*(2), 281–306.

Fleer, M. (2014). The demands and motives afforded through digital play in early childhood activity settings. *Learning, Culture and Social Interaction, 3*(3), 202–209.

Fleer, M. (2019). Scientific playworlds: A model of teaching science in play-based settings. *Research in Science Education, 49*(5), 1257–1278.

Fleer, M. (2021). Conceptual playworlds: The role of imagination in play and learning. *Early Years, 41*(4), 353–364. https://doi.org/10.1080/09575146.2018.1549024

Fleer, M. & Kennedy, A. (2000). Quality assurance: Whose quality and whose assurance?. *New Zealand Research in Early Childhood Education, 3*(2000), 13–30.

Fleer, M., Gomes, J., & March, S. (2014). Science learning affordances in preschool environments. *Australasian Journal of Early Childhood, 39*(1), 38–48.

Fleer, M., & Pramling, N. (2015). Knowledge construction in early childhood science education. In *A cultural-historical study of children learning science* (pp. 67–93). Springer.

Freeman, S. (2021). *Opening Doors. Guiding teachers to intentionally facilitate science for young children* (Doctoral dissertation, Open Access Te Herenga Waka-Victoria University of Wellington).

Hirsh-Pasek, K. (2009). A mandate for playful learning in preschool: Applying the scientific evidence. New York. Oxford University Press.

Klahr, D., & Nigam, M. (2004). The equivalence of learning paths in early science instruction: Effects of direct instruction and discovery learning. *Psychological science, 15*(10), 661–667.

MacNaughton, G. (1995). The gender factor. In Creaser, B. & Dau, E. (eds). *The anti-bias approach in early childhood*. Melbourne: Longman.

Meng-Fei, C., & Jang-Long, L. (2015). Investigating the relationship between students' views of scientific models and their development of models. *International Journal of Science Education, 37*(15), 2453–2475. https://doi.org/10.1080/09500693.2015.1082671

Ministry of Education. (1996). Te Whāriki: He whāriki mātauranga mō ngā mokopuna o Aotearoa. Early childhood curriculum. Wellington: Learning Media.

Ministry of Education. (2007). The New Zealand Curriculum. Wellington: Learning media.

Ministry of Education. (2017). *Te whāriki. He whāriki mātauranga mō ngā mokopuna o Aotearoa: Early childhood curriculum*. Retrieved from https://www.education.govt.nz/assets/Documents/Early-Childhood/ELS-Te-Whariki-Early-Childhood-Curriculum-ENG-Web.pdf

Pendergast, E., Lieberman-Betz, R., & Vail, C. (2017). Attitudes and beliefs of prekindergarten teachers toward teaching science to young children. *Early Childhood Education Journal, 45*(1), 43–52. https://doi.org/10.1007/s10643-015-0761-y

Penner, D. E. (2000). Explaining systems: Investigating middle school students' understanding of emergent phenomena. *Journal of Research in Science Teaching: The Official Journal of the National Association for Research in Science Teaching, 37*(8), 784–806.

Watson, R., Goldsworthy, A., & Wood-Robinson, V. (1999). What is not fair with investigations? *School Science Review, 80*(292), 101–6.

Windschitl, M., Thompson, J., & Braaten, M. (2008). Beyond the scientific method: Model-based inquiry as a new paradigm of preference for school science investigations. *Science education, 92*(5), 941–967.

Chapter 5
Discussion of Emerging Themes

Theory, Curriculum, and Practice

Abstract The reporting of this case study was facilitated by careful data collection and storage over the length of this research. This also meant a substantial amount of data to make sense of which was challenging and required careful analysis of each data set, meticulous checking by peers and critique by a critical friend. In this chapter, we share the themes that emerged from the integration of both inductive and deductive data analysis. Seven themes are discussed, and the chapter concludes with a summary.

Keywords Implementing our ECE curriculum · Teacher as a learner · Learning stories to learning conversations

5.1 Emerging Themes

These themes have been arranged around the intended foci of the research project, namely, curriculum, teacher development, mentoring, opportunity for children engagement, children's science learning, theory and practice, the bicultural aspect of the curriculum and assessment of learning through learning stories.

5.1.1 Understanding and Implementing the Intentions of Our World Leading Curriculum

> Bernstein's unique contribution to critical curriculum studies is to uncover ideology and power relations by examining the underlying structures of school knowledge and practice. His notion of educational code refers to a "regulative principle, which underlies various message systems, especially curriculum and pedagogy" (Atkinson, 1985, p. 136).

With experts designing the curriculum, teachers implementing it, and the New Zealand Education Review Office evaluating the implementation, there is diversity of interpretation, and this risks a lack of shared understanding. This is not unique to this curriculum, but more common to other curricula globally. In the 1970s, Bernstein theorised that distribution and packaging of educational knowledge, along with its arrangement of curricula and selection of disciplinary classification advantaged those who were within the hierarchy of the structures. The changes that followed in the next five decades rejected these disciplinary approaches and prioritised integrating curricular knowledge across all school disciplines (Pluim et al., 2020). The metaphor of te whāriki, the woven mat, that underpinned the curriculum design of our ECE curriculum was an early response to this notion of curriculum integration. Te Whāriki (Ministry of Education, 1996) incapsulated all the experiences, activities that take place within the learning environment that nurtures children's learning and development. The curriculum aspired the children to:

> to grow up as competent and confident learners and communicators, healthy in mind, body, and spirit, secure in their sense of belonging and in the knowledge that they make a valued contribution to society. (Ministry of Education, 1993, p. 9)

To acknowledge New Zealand's bicultural constitution and to be true to the partnership between Māori, the indigenous people, and the Pākehā European settlers, the curriculum reflected the partnership in its principles. The Treaty of Waitangi is the foundational document of this country and has three broad agreements for participation, protection, and partnership. It is "a bicultural early childhood curriculum that validates and enacts kaupapa Māori" (Richie, 2005, p. 109). The guiding principles of the curriculum were and continue to be empowerment of all children, holistic approach to development and learning, acknowledging the role of family, and community and importantly, relationships between all involved in the children's lives. The curriculum had 5 strands: Wellbeing, Belonging, Contribution, Communication and Exploration. The curriculum was bilingual, and its integration was represented as the woven mat symbolised in Fig. 5.1.

As the curriculum design was based on a set of principles and did not provide a structure for planning, even at the draft stage Cullen (1995) highlighted the tension caused by the weakness in the knowledgebase and training of ECE teachers. So, the curriculum was aspirational with high ideals but how the intent of the curriculum would be embraced by those who were to implement it, the educators, and those responsible for ensuring its implementation, the Education Review Office were not so clear. Bridging this gap between the intent and implementation was challenging for both policy makers and teachers. Over the years of its implementation the education of the everchanging nature of the workforce remained on the acknowledgement of biculturalism, but the implementation prioritised child development and care and not learning.

In the first part of the present research, an interesting dilemma for the teacher was the lack of synergy between what she learnt in her teacher education and to comply with the focus of the centre in which she was employed. The idea latched on to by the centre was to *follow the child's interest*. The teacher felt constrained when providing science related exploration activities. This was pointed out to her by her employer that she was not following the child's interest but trying to teach! It took a while and persistence on her part for her colleagues to understand that

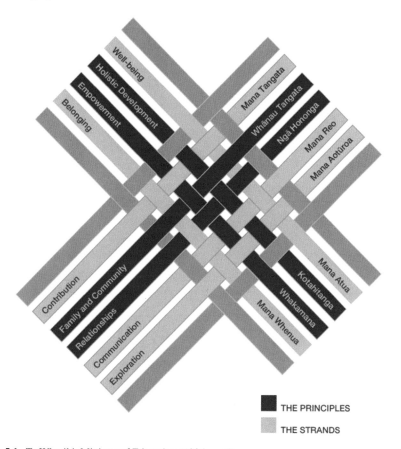

Fig. 5.1 Te Whāriki, Ministry of Education (1996, p. 13)

the more exploration opportunities they provided, the more likely the children were to show their interest beyond the iconic, painting, collage, playdough, sandpit, and riding tricycles. It took a while to get across that when children were exploring fruit, they were interested in them AND at the same time learnt how to look at the fruit closely. Such activities were rich opportunities for sensory exploration, developing oral language, counting, and ideal opportunities to introduce describing, colour, texture, smell, and taste both in te reo Māori and in English, as intended by the curriculum. Modelling that children can only show their interest in things that they get to see, feel, and know required embedding in the centre's culture. In one centre where she was employed, the Education Review Office wanted to know what the purpose of the fruit of the week's activity was? Although present, the teacher was not asked to respond, instead the manager said, 'it was good for the social development of the child'.

Blaiklock (2010) pointed out that the integrated and holistic nature of the curriculum resulted in subject content areas (e.g., art, music, science, literacy) being overlooked. Further, that the generalised nature of the guidelines on planning may be leading to inadequate range of learning experiences. Blaiklock was also concerned about the value of Learning Stories, that were a novel assessment approach designed to align with the approach of Te Whāriki. We discuss learning stories later in another theme in some detail. In the decade that followed, New Zealand society has become increasingly multicultural due to migration in the last 30 years.

In updating Te Whāriki in 2017 while remaining committed to the bicultural nature of our nation, the focus has changed to include others.

> This curriculum acknowledges that all children have rights to protection and promotion of their health and wellbeing, to equitable access to learning opportunities, to recognition of their language, culture, and identity and, increasingly, to agency in their own lives. These rights align closely with the concept of mana. (Ministry of Education, 2017)

Identity, culture, and language now have primacy.

> Learner identity is enhanced when children's home languages and cultures are valued in educational settings and when kaiako are responsive to their cultural ways of knowing and being. For Māori this means kaiako need understanding of a world view that emphasises the child's whakapapa connection to Māori. (Ministry of Education, 2017, p. 12)

The other feature of the updated curriculum is the change of emphasis on learning alongside care and development, which continue to be underpinned by the original principles. Guidance is given on how the curriculum structure is organised and learning outcomes are provided. "The learning outcomes are broad statements of valued learning. They are designed to inform curriculum planning and evaluation and to support the assessment of children's progress" (Ministry of Education, 2017, p. 16).

In the beginning, the mentor needed to help Seema unpack the intention of the curriculum and explain how by talking about fruit and sensory exploration she was integrating physical and language development. Encouraging the children to eat the fruit was good integration of wellbeing habits. The group exploration of fruit was excellent opportunity for social development, to say a karakia (a traditional prayer), and taking turns. That guidance was enough to get her to think about the potential of the exploration opportunities she was providing and soon she was able to successfully look for integration opportunities.

The 2017 changes are considerable where the shift is from a curriculum that did not provide a framework for planning to a far more structured approach with guidance about how the change could be managed. An hour-long discussion using the Overview of the Te Whāriki, Strands, Goals, and Learning outcomes (Ministry of Education, 2017) took place between the mentor and the teacher to unpack the changes. Figure 5.2 provides the overview for the Exploration strand, which is relevant to this research. Similar guidelines are provided for all other strands which were unpacked with her using examples.

The change to the curriculum and the meetings she had at the Centre to come to grips with what changes to the curriculum meant for pedagogy, almost caused her to panic. She needed guidance to unpack each strand comparing her current practice and

STRAND	GOALS	LEARNING OUTCOMES
Exploration Mana aotūroa	Children experience an environment where:	Over time and with guidance and encouragement, children become increasingly capable of:
	» Their play is valued as meaningful learning and the importance of spontaneous play is recognised	» Playing, imagining, inventing and experimenting | te whakaaro me te tūhurahura i te pūtaiao
	» They gain confidence in and control of their bodies	» Moving confidently and challenging themselves physically | te wero ā-tinana
	» They learn strategies for active exploration, thinking and reasoning	» Using a range of strategies for reasoning and problem solving | te hīraurau hopanga
	» They develop working theories for making sense of the natural, social, physical and material worlds	» Making sense of their worlds by generating and refining working theories | te rangahau me te mātauranga

Fig. 5.2 Overview of the Te Whāriki, strands, goals, and learning outcomes (Ministry of Education, 2017, p. 25)

what the changes might look like. The updated curriculum change had made a significant change for Seema's pedagogical practice. The curriculum structure provided the framework- strands, goals, learning outcomes and what achievement of these learning outcomes would look like. It allows for intentional planning, encourages literacy development both for English, Te Reo Māori, as well as Pasifika languages. This curriculum provides teaching strategies and resources that are clearly set out. Similarly, the rationale and ideas for assessment for learning are provided. Seema has been a committed teacher in trying to deliver the bicultural curriculum. She had continued to learn the languages herself and seek opportunities to integrate them. In the last year of this research, she began to teach the sign language to her children as it is New Zealand's third official language.

Our observation from Seema's management of this transition to the new curriculum has made us wonder why our country that puts in significant resources to develop such a world leading and inspirational curriculum then leaves it to individual centres to implement it. With a range of ECE centres, some with minimal trained staff and others with more, we do need to have quality professional development that helps teachers understand the intent and give opportunity to practice the pedagogical changes with guidance. Our research shows that an ECE teacher who is at the centre from 8 to 5 does not have the energy to sit through two hours of meeting and professional development (PD) at the end of the working day. For sure, this practice ticks the box that PD has been offered and taken, but does it support the implementation of the intended curriculum?

Similarly, expecting teachers to read and understand the new curriculum, its structures, requirement, and using the resources in their own time is unrealistic expectation for someone who has already worked a long day which requires physical exertion.

Teachers are already planning and reporting progress in their own time. Putting the resources on the website is not sufficient if the policy makers earnestly want change to take place.

A curriculum may be aspirational, and we think it should be, but it also needs to be achievable. Doing research to find out that the teachers did not implement the curriculum as intended is a waste of time and resources. If writing an outstanding and aspirational curriculum is worth spending resources on, ensuring that we enable teachers to implement it with guidance is critical and worthy of the investment.

5.1.2 Teacher as a Learner—The Key to Professional Development

An impressive aspect of Seema's approach to her own learning was that she wanted to learn and develop ways of using her new learning to provide learning opportunities for the children in her care. She not only asked for guidance from the mentor but very early on started to notice how children were learning from the activities she was organising for them to play and explore. She did not offer excuses as to why she could not do something and tried to have a go at almost anything suggested or those that she thought of herself.

She notices. This was one of her strengths, she senses when the children were likely to engage in an outdoor activity or how while the children were busy playing in the sandpit, introducing water play to extend their learning would be useful. She noticed things in the surroundings and naturally pointed these out to the children. On one such occasion when the children were outdoors, she gave them the telescope she had brought from home. She was very excited that the children worked out how to use it correctly and joyous on reporting that Tama told her that 'a telescope is used to see things that are far away'. Her follow up activity was to put out the magnifying glasses and some insects for children to see that it is possible to use magnifying tools to make things close bigger, just as a telescope makes far away things closer.

She was brave to try any feasible exploration activity that was safe to do. For example, her children made earthen lamps out of clay. She asked children how they could make them stronger and cooked them on high heat in the oven! Then the children used it to fill it with some oil, making a cottonwool wick and lighting the lamps to celebrate Diwali, the Indian festival of lights. Along with having a go at all sorts of science activities, she developed the habit of trying things out at home before putting them out for children to explore. Another example was when she was going to add baking soda to water in a glass and add some rice to make it dance. She said that she tried several times and worked out that if she broke the rice into smaller pieces, little bubbles carried it to the top and made it dance. When she set the activity up for the children and they were having fun adding broken rice and seeing it dance, she gave them some large grains of rice. They did not dance. The children had to come up with their working theory why the large grains sank. Her excitement and sheer joy when sharing this with the mentor was wonderful as one child had said that it was too big, and another suggested it was too heavy.

The teacher sought advice and acted upon it. She did not regard the mentor's questions as criticism, but as opportunity to improve her practice. Going over the interview data and mentoring notes, her enthusiasm to try new things and a sense of achievement when a child learnt something new was shared with other teachers and parents.

Three things that she found challenging. One, when she was given a pile of documentation to go through and work out how to put them into practice. Two, curriculum changes, and three, the constant change in the management of the centres in which she worked. The last seemed to be very common in the last two centres. The pile of documentation were challenging because they were emailed to the teachers and with that instructions to the changes that they would need to make. Often, this was in response to either the eminent visit from the Education Review Office (ERO), or after they had visited. It was that she was complying with what was required as pressure was put on teachers to prepare for the ERO visit. It was almost as though all teaching and learning was put on hold while every detail of the previous ERO report was complied with. This often resulted in a neglect of other aspects of care and learning and a clear focus upon the 'new things to improve' before the next ERO visit. It is important to mention here that none of these required changes in the three centres she worked were due to unsafe practices or lack of care. Nevertheless, the eminent visits were always a stressful time. ERO (2020) have proposed a forward-looking developmental approach to evaluation where they will support each school's improvement over time. This will also have a collaborative approach where and ERO evaluation partner will work alongside each school. Seema will now encounter a new way of working with ERO and with-it new opportunities and challenges.

The reaction to curriculum changes was extraordinary. Perhaps, the changes was being implemented to the current understanding of the centre management who were also coming to grips with it. Donaldson (2015) suggested that when teachers are involved in co-constructing change and experience agency the outcomes are positive and empowering. Conversely, when teachers are recipients of top-down change, and asked to passively deliver policy then outcomes are less promising. Centrality of teacher agency is influential in improved learner outcomes (Harris & Jones, 2019). The change from Te Whāriki (1996) to Te Whāriki (2017) was significant and our view is that the uncertainty and stress was caused by trying to make several changes all at once with inexperienced centre leaders.

Changes in the management cannot be helped. In her last centre she has had three head teachers and three new managers in the space of less than two years. Each time either the manager or the head teacher changed, the new person wanted to forget about anything that was there and start with their view of how the centre was to be run. The inexperience and lack of understanding on the part of the leadership led to direct instructions about what needed to be done. Seema found this change from good practice, which was working for the children to something new that was not deeply thought through unsettling and challenging. Here the mentor's support and reassurance that she was doing the right thing was needed.

She took every opportunity for professional development offered for implementing the curriculum, learning te reo Māori, conducting inquiry, attending first aide training and visits to other centre to learn how they did things. In accordance with Skerrett and Ritchie's view (2021), Seema like many others in New Zealand demonstrates the need to be cognizant of Māori parent's desire for their children to learn their language and continues to seek opportunities to improve her own knowledge and to ensure children are learning it alongside English. She organised and managed fieldtrips, had learning conversations with the children and documented them, learnt from parents, and shared the successes of their children with their whānau. Evidence suggest that she has become a competent carer, ECE educator and science teacher of small children. She values play but also ensures that children are learning science through play.

5.1.3   Mentor as Both Teacher and Learner

As the data were analysed it became clear that this journey of learning was for both the teacher and the mentor. Learning to listen and getting better at it helped the mentor to find out how Seema learnt. For example, initially she expected the mentor to tell her what to do and explain why. This meant a lot more talking on the part of the mentor. The diary entries suggested that with time, there was little that the mentor had to say, but to listen and affirm what the teacher was thinking. Or to ask a question to guide Seema towards what to try out and what might work.

Relationship building and sustaining this relationship over time made it possible to continue the learning journey. The mentor had learnt to delegate responsibilities to other members of her team earlier in her career as teacher in charge of a large science department. Knowing to trust, encourage, and support when things did not work was a useful skill she brought to this mentoring process. The mentor believed that maintaining a balance between guidance and taking over was essential for the relationship. Seema knew that she could ask for help but also that as her confidence grew, she needed less and less help. Nolan and Molla (2018) recommend "mentoring to be a socially situated practice rather than a detached pedagogic event" (p. 551). Being cognisant of the ECE environments, relationships and expectations was important to avoid cognitive dissonance between the guidance provided by the mentor and her centre's expectations.

When faced with a pile of documents to get her head around, the mentor suggested that Seema read just one of the documents at a time, taking time to consider its meaning at her own pace. Think about what she was doing at present and what the change was asking? If change was needed, then what would Seema do herself. In the first three years, Seema needed help to understand the new jargon in the document. By year six, she was able to figure it out herself. Her strategy of breaking down the new material into smaller chunks to read and understand turned out to be a very useful strategy. Part of the panic was the sheer volume of the reading and sense making to be done. Once reduced to bite sizes, it became easy to manage. By year 9, Seema had settled into a clear routine of downloading one document at a time!

It appears that the mentor kept two foci in sight, focussing on enriching teacher practice and enhancing children's learning. Acknowledging and celebrating the sheer excitement that became increasingly obvious in the learning stories made it visible to Seema that the mentor is genuinely interested in her planning and impressed by what the children were learning. That said, the mentor had to encourage Seema to create multiple play and learning opportunities before the children understood the underlying science idea. Doing something once, may lead to situational interest and some learning but reinforcing is useful for long-term learning (Moeed, 2016; Palmer, 2009).

In Chap. 3, we had highlighted some questions that arose from mentoring, here we unpack these questions:

1. *How effective are the current practices of teacher professional development?*

Based on the professional development opportunities Seema was provided, the PD sessions at the end of the day after a full day of working were least useful. It could have been the teacher's ability to concentrate or sitting and listening to someone presenting new ideas. It apparently did tick the box when it came to participation in PD. But throughout the study, it appears to be the least effective. In her present ECE centre they focus on one document or change idea at a time with opportunities to try out and discuss before moving on to the next document, this has been more manageable. In Australia, Colmer et al. (2015) report that although centre directors said they valued collaborative professional learning, the one-off external PD events were the common reality. That appeared to be the case in the first 5 years of this study. Seema gained more from the two-day weekend PD that she had chosen to participate in because it was an area she had identified as her weakness and needing updating and although it was an entire weekend, she felt she could concentrate and take things onboard.

2. *Why is there a compulsive need to add new terminology?*

ECE teachers are expected to communicate to the parents and whānau, what they have been doing to care for and support their children's learning. Much of the vocabulary used in ECE is not accessible to people not involved in teaching. Reporting to parents about their child's disposition, schema, or the strategies used is less helpful. Rather than talking about the child's disposition explaining that the teacher has noticed the child's attitude towards learning activities is far more useful. There is no shortage of jargon when it comes to teaching strategies. Using words like metacognition, may sound impressive but teaching the child to think about their thinking is easier to understand. All that is being said is we are asking the child, why do you think so? That is what is needed for the child to be meta cognitive. Similarly, what does 'developing more complex working theories' mean and look like in this context. Talking about serve and return as a teaching strategy could simply be asking question to continue a conversation.

Our curriculum uses many Te Reo Māori words, which is a good thing in a bicultural country. However, Māori words are not just words, there are underlying principles which need to be understood. For example:

Empowerment (Whakamana)

> This principle means that every child will experience an empowering curriculum that recognises and enhances their mana and supports them to enhance the mana of others. Viewed from a Māori perspective, all children are born with mana inherited from their tīpuna. Mana is the power of being and must be upheld and enhanced (p. 18) (Emphasis is added to highlight the Te Reo Māori words).

To be fair, this is unpacked in the following paragraph. But further down on the same page the curriculum says:

> Perspectives on empowerment are culturally located, hence kaiako need to seek the input of children and their parents and whānau when designing the local curriculum (p. 18).

There is potential for misunderstanding any or all these words unless unpacked to ensure shared understanding for all. New Zealand has a very diverse teaching workforce, some trained and others not. This requires effort on the part of the management to make sure how these will be unpacked and understood.

3. *What teacher support is provided when transitioning into a new curriculum that has significant changes?*

In New Zealand, teachers are challenged to bridge the gap between the overall principles and strands in Te Whāriki and their actual pedagogical practices (Alvestad et al., 2009). Teacher support in transitioning into a new curriculum appears to be up to individual ECE centres. The quality of support depends on the size of the centre, leadership as well as teaching workforce. Cherrington and Thorton (2013) identified factors that affect the professional learning and development these include funding models, teachers' workloads, whether relief teachers are available as well as leadership (Cherrington & Thornton, 2013).

4. *How effective is putting information on the Ministry of Education website for teachers to read and come to a shared understanding for creating a local curriculum?*

Te Whāriki mandates an aspirational, individualized, and responsive curriculum to be tailored for each unique setting. Further, it promotes the notion of a local curriculum, that meets the aspirations of the families whose children attend a particular centre (Aspden et al., 2021). Yet evaluation reviews by the ERO highlight concerns about effective curriculum implementation and the learning experiences offered at ECE centres (Education Review Office, 2015).

This research has raised our awareness that providing the curriculum, teaching resources and support materials on the website was challenging for the participating teacher. It was much better when in her third ECE centre where they collectively talked through a particular resource at the once a week after school meeting. However, with the limited training and understanding of the leadership it became constraining rather than enabling. One such example was when it was suggested that the teachers put puppets as a learning strategy in the children's ILP (Individual Learning Plans) to encourage story telling for oral language development. The insistence that puppets MUST be used became a challenge for Seema when the children were not interested in playing with them. The leadership insisted, she uses them because it is in the children's ILP!

Finding the time to read what was on the website, know the interest of the children, their whānau's aspirations and choosing appropriate strategies in our view is an unrealistic expectation perhaps achievable by some. If the intension is for the curriculum to be implemented as desired better mentoring is required.

5. *In what ways would timely mentor support be better than collective professional development sessions?*

We have found that for quantity of professional development was not as helpful as timely, clarification and support from the mentor. Perhaps a model like School Support Services that provided advisors would be a better option. School Support Services in New Zealand were an arm of teacher education for in-service teachers. If there was a change in the educational policy, they offered group sessions to take teachers through what the documentation was asking. Then if needed, schools could access them for individual support as and when required. The advisory services were abolished in the early 2000s and replaced with a government funded PD model provided by individual, approved service providers. The priority of what will be supported is decided by the government annually. Considerable time is invested by private organisations to secure contracts. On the one hand, this may be a more agile model of PD delivery able to quickly adjust to changes in government policies and priorities. However, it comes at a cost as longer-term relationships between users and provider are likely sacrificed for a market driven PD model.

Currently, a support system as the idea of "communities of learning" which are made up of clusters of schools supporting each other is in place. There is little research on the effectiveness of such networks. However, recent research indicates that although networks within the schools are connected, there is low level of advice seeking and school collaboration across schools (Sinnema et al., 2020). Perhaps, having an identified mentor who can support the teacher would work. We suggest that if an ECE teacher is wanting to develop skills for teaching science, they could gain from being paired with a mentor teacher from the local initial teacher education provider who they can approach as and when they need support.

5.1.4 Exploration, Playing, and Learning Science Through Investigation

Analysis of the data showed the following sequence of presentation of learning opportunities (Fig. 5.3).

Fig. 5.3 Sequence of
practice of presenting
learning opportunities

Starting from providing opportunities to play and following the child's interest, with experience Seema considered the learning experiences that would provide opportunities to explore. Gradually, she began to organise exploration of materials that were not so easily accessible to all children. This does not mean that play was replaced by a structure, rather play was encouraged and opportunities were provided to explore more. In the early years, these included flowers, leaves, and water play. It is through these explorations as children played and interacted with material, she introduced science ideas such as floating and sinking. While playing, children engaged in simple problem-solving activities working out what things float and sink. Allowing children to explore fruit and use their senses to make observations.

She appeared to encourage play that reflected Fleer's (2021) suggestion that "imagination in play is foundational for imagination in conceptual development" (p. 353). For example, when children played with magnets through questioning and encouraging children to try out possibilities, she facilitated the activity adding resources and information in such a way that they were able to gradually work out that when the same sides of magnets are brought together, they push away while others make them stick together. This would later become the foundational understanding of the science concept of like poles of magnets repel and unlike poles attract each other.

Deliberate teaching of investigation started when the new revised curriculum encouraged children to come up with their own working theories. Working theories had to be explained to her as a concept. Children grew sunflower plants from seed and when one of the younger children pulled out a plant, she confidently used it as an opportunity for the children to learn about the different parts of the plant rather than reacting and showing disappointment.

There were several examples where the teacher quietly observed children playing and provided more materials so that they could extend their play. This was common in children's outdoor play when she was supervising. Providing a hose, some digging tools, balls or even hoola hoops on a fine day added to the children's play, exploration and trying out more challenging physical moves. Similarly, setting up outdoor equipment in different configurations so that the children climbed, slided, and took on the challenge to walk along a plank. Some children needed a supporting hand to gain the confidence to try out something they found challenging. Such play encouraged communication, fostered imagination, and promoted creativity (Vellopoulou & Papandreous, 2019).

An example of play leading to investigation was when children were playing with ice blocks. One child did not want his ice block to melt and buried it into the sandpit. Later he took Seema to have a look and dug up his ice block. He informed her that he had put some water in to keep it cool and frozen and that he thought that sand was an insulator! This activity showed that the child had encountered, tried out and begun to develop an understanding of the concept of insulation as suggested by Bodrova and Leong (2015). Fleer (2015) suggests that such learning through play occurs through thoughtful teacher involvement. It appears, that engaging in play alongside the children brought joy to Seema who often excitedly reported what the children were saying or thinking in her learning stories.

Through rich play like making fruit salad after the children had been learning about different fruits and making it into a tasting game was a lot of fun providing opportunities for discrete science; embedded science and counter intuitive science when the children could not identify a fruit while being blind folded! This approach is encouraged by Sikder and Fleer (2015). In sum, children played, explored, and investigated.

5.1.5 Ways of Making Sense of Our Natural World Through the Adoption of Mātauranga Māori, Science, and Other Perspectives

One purpose of science education is to help children understand the natural and physical world. Historically, it was believed that preschool children's understanding of science concepts was not investigable because children could not understand science phenomenon (de Kock, 2005). Recent brain development research shows that the human brain develops rapidly in the first few years.

de Kock (2005) explains that children begin to grasp ideas very early, for example, when they use the see-saw, they can see that by pushing up one end of the see-saw, the other end will go down. They may not be able to articulate their understanding in the terms used in physics. In the present study, there are several investigations through which children could explain the phenomenon of floating and sinking, melting, and freezing, magnetic force, push and pull force, and the needs of small animals such as snails and butterflies. These understandings were developed by the rich learning experiences and opportunities to play provided by the teacher.

It is important to recognise that all human children explore their surroundings and make sense of their world. In all human communities, there are understandings that the children's development are mediated by the culture. For example, in the Māori culture there is tikanga for washing hands before eating food. This really is the idea of safe eating practice to not pick up infections or bugs going around. Muslim children are trained to wash their hands three times before eating. Similarly, in recent research conducted at a Māori medium school, children learnt that the monarch butterfly can find a swan plant because the butterfly whakapapa's to the wind (Whakapapa is the genealogical connection). The smell must be carried by the wind and the monarch butterfly can smell the plant even when it is not in flower! (Moeed & Cassidy, 2020).

Mātauranga Māori is indigenous knowledge in New Zealand and Mercier et al. (2019) argue that in "Māori Science, students examine Mātauranga (Māori knowledge), its scientific nature and its connection with Western science" (p. 140). For children in ECE, this can begin with learning the Te Reo Māori (the Māori language) and the tikanga which is applicable in everyday life. Finally, and importantly, our society is becoming increasingly multicultural and therefore being cognisant of other ways of knowing and acknowledging them is important.

5.1.6 Theories and Practices; Communitas, Socio-cultural Theory and Social Constructivism

In this complex yet illuminating case study we have focussed on three theoretical approaches. Communitas is an emotional experience which is spontaneous, immediate, and concrete (Turner, 1969). In the ECE situation there were many examples of spontaneous experiences, when a butterfly flew into the classroom, when someone poured water from a small cup and it took a long time to fill a jug, when children noticed that the seeds, they had been watering had germinated. In the ECE classrooms this teacher worked in, there were very few structures that constrained play and learning. In fact, until the new curriculum suggested deliberate acts of teaching, all interactions were spontaneous. There was obvious joy when something worked or not when it did not work. There were a lot of cheers when a large ice balloon the children thought 'would **definitely** drown, decided to float'! There were other opportunities for the children to push and pull the tables and workout what was easier.

In several stories, the sheer joy that the teacher experienced came through. Like when a boy brought along a snail and waited for her to arrive and later when he told her that his snail had gone for a walk and came back with a friend.

Life in an ECE centre is flexible it is a place where many emotions are shared, and many relationships built based on emotions and without structures (Wilmes, 2021). Evidence of collective emotion when the children were caring about little animals or growing plants. The sheer joy of tasting some fruit they had never eaten before or learning to do a gymnastic move that they had seen a friend make.

ECE centres are communities of learning where, teachers, parents, members of whānau for example grandparents, and other children work together all day. In such communities everyone shares the responsibility of children's care, learning and development. Rameka et al. (2021) and others have agreed that 'it takes a village to bring up a child'. Communitas underpins this thinking.

In New Zealand a government initiative, encouraged A Nation of Curious Minds—He Whenua Hihiri i te Mahara was established (New Zealand Government, 2014). It involved collaborative partnerships between communities and scientists with a view of raising scientific literacy and engaging public in science that was taking place in the community. This initiative funded several projects that involved ECE centres particularly in environmental education projects. One such example is the National Institute of Water and Atmospheric (NIWA) research project in South Auckland where scientists and children investigated the air quality in their region. South Auckland has many factories and very high incidence on respiratory illnesses among young children. An interesting collaboration with positive outcomes for the community.

Human babies are not born socially competent, much of what they learn is through social interactions. Sociocultural theory argues that children learn from social interactions in the society (Rogoff, 2003). Seema noticed that when children arrived in the centre they were focussed on self and their own needs. There are several occasions in this research when children were reminded to share things or take turns. Waiting patiently, while someone else has a turn is a learnt skill that the children developed while playing together. One interesting idea shared by the teacher in a meeting was that all the children could tell you what the Kawa (rules) of the centre were. Slow and quiet feet, no pushing, no throwing things. However, they did not practice these! She said she had to model such behaviours and remind them. Knowing what the right thing was not enough. Behaviour needed to be practised, acknowledged, praised, or reminded depending on the child. An added complication in ECE centres is that children turn five and leave and new ones arrive, so acceptable behaviour needs to be reinforced repeatedly.

Veresov (2017) proposed that concept of social relations, and social reality are the foundation of development, and explains that development is the process of the individual becoming a social being. There were many examples of children learning with and from each other. Once the children worked out that Sam was interested in little animals, as soon as they saw an insect or spider, they would call him to come and have a look. Soon they acknowledged him as the expert who could identify these. Children learnt Māori tikanga (customs) from the teachers and others. As a result of teacher modelling, they learnt the Māori greetings and began to use it when they arrived and left the centre. Similarly, through listening to the story about Tane Mahuta (The largest Kauri tree in New Zealand) children became interested to make a poster together. Sociocultural theories emphasise the involvement of learners in the social practices within their unique context (Danish & Gresalfi, 2018). Celebrating the Māori, Samoan or Tongan language weeks provided ideal opportunities for children to learn about the social practices of these cultures. Similarly, the teachers organising a Diwali celebration created space for children to make earthen lamps for light. The related activities became more involved over the years.

The curriculum acknowledges and celebrates the diverse cultures within this country and there were several opportunities for the teacher and the children to learn from the children's families about their cultural practices. To gain knowledge about Māori culture the teacher attended three days of professional development at a marae (Māori meeting place) in the Hutt Valley. There she learnt the significance of a korowai (a Māori cloak made of feathers) and how it was made. After gaining knowledge about korowai and the tikanga followed in making them she invited a mother who had the experience of making korowai to show the children a real korowai. The mother helped the children make a paper korowai. The children were exploring feathers, and this was followed by them cutting out paper feathers, colouring them and together making a korowai which is traditionally made with feathers. Children loved taking turns to put it on and have their pictures taken.

The third theory used in this research was social constructivist theory. Constructivist theory prioritises finding out what the learner already knows and presenting new information in ways that help them to make sense of it. After year 3, most activities organised by Seema had a constructivist approach. In her reflections she wrote about what they had done, what she thought the children had learnt, and what she would plan to reinforce or extend their learning. Constructivists theories have been prioritised by science education in New Zealand where it is believed that knowledge is individually constructed but socially mediated (Moeed, 2016). Examples from the teachers practice include, children making observations of snails, then talking about what they eat and investigating whether they preferred to eat cabbage or lettuce leaves. When the children said the snails preferred lettuce but also ate the cabbage, their teacher reminded them how they eat the food they like best first from their dinner plate and then mum encourages them to eat their less favoured foods!

Evidence suggests that all three theories that underpinned this research were evident when the data were critically analysed.

5.1.7 *Learning Stories to Learning Conversations*

Learning stories have been adopted and used extensively to assess children's learning processes, learning dispositions, and well-being. Blaiklock (2008) raised concerns about learning stories which included issues to do with validity and accountability of learning stories, teachers' decisions might be made on short observations and may be subjective. He was also concerned that teachers may not have adequate guidance on when and how often to write learning stories. Further concerns included assessment of children's dispositions and in reporting progress over time. Blaiklock also believes that teachers might be using the assessment information as they respond to the children at the time, but he found little evidence to suggest this was being used for future planning. This was also the case in the first two years of Seema's teaching. She was able to identify children's learning need and responded to it but as she was still grappling with science herself, she needed mentor advice to encourage her to plan for future learning. In the last five years, this had changed, she could spot what the child's need was and come up with a creative activity to challenge them to think. For example, when the children had been exploring magnets, she brought along a toy that required children to apply what they had learnt, in their exploration into a new context as they played with the various components of the toy system. Children worked out how to move the toy cars without the magnet touching them. They also worked out that they could make the little toy people jump by moving the discs.

The mentor recorded a conversation where she asked Seema to talk about a child's portfolio. She said:

Portfolios have pictures, stories, and children's work. They are kept at the centre Parents and children can see these. Children often look at these and can see the progress they have made (Interview Year 3).

The mentor asked her, "how can children see their progress by looking at their portfolio"? She stopped and thoughtfully answered that they can see in the pictures that they have grown. However, they cannot really read what the learning story said. This was the turning point when they discussed considering changing the format of the learning stories. She could see that writing a story which was written from entirely the teacher observation was not as useful as having a conversation with the child and documenting it. This led to the shift to writing learning conversations as exemplified in several instances in Chap. 4.

The main difference between learning stories and learning conversation is the shift from being teacher-centred to child-centred approach to assessment. Instead of the teacher just writing what they saw and what they thought, Seema began to talk with the children to find out what they had created, encouraged them to talk about it and about what they were thinking. This she noted created the opportunity to introduce new words and helped in oral language development. In the last year it is noteworthy that she began to ask them, why do you think so? This was a good start to encouraging the children to be metacognitive.

A major limitation of learning stories is that they do not deliver on the 'construction of learner identity' according to Zhang (2017) which was a goal according to Carr and Lee (2012). Zhang believes that there might be two reasons for this. First, that the learning stories lack depth because it is not an easy job and is time-consuming. Second, that the data collected in the form of learning stories is not synthesised to draw valid conclusions. With the change in Te Whāriki (2017) which emphasises the learning outcomes to be set and met, a change in practice is likely that will counter and respond to these weaknesses. For example, we have noted a difference at Seema's ECE centre where they write each child's individual learning plan for each half year. They identify the strand of the curriculum they are focussing on and describe the learning outcomes. They also identify the strategies they will be using to support the children to meet these outcomes. Importantly, at the end of the half year the teacher reviews all the learning stories and synthesises the information to write and evaluation report. New goals are set and the learning outcomes they will be aiming for and the strategies they are likely to use are shared with the parents who offer their comments and whether they would like specific things included. For example, one parent asked for the child to be supported to develop emotional resilience. This kind of assessment is a great way forward; however, we cannot include examples of it in this research as it falls beyond the period of this project.

5.2 Summary

We began this chapter by explaining that seven major themes had emerged when the data from various sources were analysed. These sources for these themes included teacher interviews, teacher reflections, mentor notes, and children's learning stories. The themes related to the different aspects of this long-term investigation. The themes were discussed in the light of recent, relevant literature. We explained the goals and expectations of our curriculum and how they changed with the review of the curriculum. The focus moved from reporting progress towards the goals to reporting the on the set learning outcomes. We found that it was critical that the teacher has a 'learner mindset' that actively engages with where the individual child is in their learning journey, and this entails supporting the child when met by challenges.

We also found that a lack of timely support in unpacking significant changes to the documentation is stressful even for a willing teacher. Professional development and what worked and did not work was commented upon. In this project there was mutual learning for both the teacher and the mentor, and this reciprocity was highlighted and discussed. Evidence of children, exploring, playing, and learning through science investigation was discussed next. The teacher learnt the relevant science as she identified the science ideas in the children's play. The major shift was in providing rich opportunities to explore followed by making the science learning visible through intentional planning and teaching.

We discussed the acknowledgement by the teacher that there are many ways of understanding the natural world and that in our country Māori ways of understanding the world as just as important as learning science. Here we also discussed that to understand the world in Māori ways, it is important to know the language and customs and that was the starting point in an ECE centre. We discussed how the theories that underpinned this research were evident in the data and discussed this. Finally, we focussed on assessment of learning, discussed how learning stories became richer over time and found that learning conversations were better for reporting children's progress.

References

Alvestad, M., Duncan, J., & Berge, A. (2009). New Zealand ECE teachers talk about Te Whāriki. *New Zealand Journal of Teachers' Work, 6*(1), 1–19.

Aspden, K., McLaughlin, T., & Clarke, L. (2021). Specialized pedagogical approaches to enhance quality for infants and toddlers in ECE: Some thoughts from Aotearoa New Zealand. In *Quality improvement in early childhood education* (pp. 1–19). Palgrave Macmillan.

Atkinson, P. (1985). Language, structure and reproduction: An introduction to the sociology of Basil Bernstein. Londres, Methuen.

Blaiklock, K. (2010). *Te Whāriki*, the New Zealand early childhood curriculum: Is it effective? *International Journal of Early Years Education, 18*(3), 201–212. https://doi.org/10.1080/096 69760.2010.521296

Blaiklock, K. E. (2008). A critique of the use of learning stories to assess the learning dispositions of young children. *New Zealand Research in Early Childhood Education, 11*, 77–87.

Bodrova, E., & Leong, D. J. (2015). Standing "a head taller than himself." *The Handbook of the Study of Play, 2,* 203.

Cherrington, S., & Thornton, K. (2013). Continuing professional development in early childhood education in New Zealand. *Early Years, 33*(2), 119–132. https://doi.org/10.1080/09575146.2013.763770

Colmer, K., Waniganayake, M., & Field, L. (2015). Implementing curriculum reform: Insights into how Australian early childhood directors view professional development and learning. *Professional Development in Education, 41*(2), 203–221.

Cullen, J. (1995). *The challenge of "Te Whaariki" for future developments in early childhood education. Report.* Faculty of Education Massey University.

de Kock, J. (2005). *Science in early childhood.* ACE papers, Issue 16: Approaches to Domain Knowledge in Early Childhood Pedagogy, Paper 9. https://researchspace.auckland.ac.nz/docs/uoa-docs/rights.htm

Donaldson, G. (2015). *Successful futures: Independent review of curriculum and assessment arrangements in Wales.* Welsh Government.

Education Review Office. (2015). *Infants and toddlers: Competent and confident communicators and explorers.* https://www.ero.govt.nz/publications/infants-and-toddlers-competent-and-confident-communicators-and-explorers/

Education Review Office. (2020). *New operating model, your go-to-guide.* Retrieved on December 14, 2022 from New schools Operating Model Your Go-to-Guide.pdf (ero.govt.nz)

Fleer, M. (2015). Pedagogical positioning in play–teachers being inside and outside of children's imaginary play. *Early Child Development and Care, 185*(11–12), 1801–1814.

Fleer, M. (2021). Conceptual playworlds: The role of imagination in play and learning. *Early Years, 41*(4), 353–364. https://doi.org/10.1080/09575146.2018.1549024

Harris, A., & Jones, M. (2019). Teacher leadership and educational change. *School Leadership and Management, 39*(2), 123–126.

Mercier, O. R., King Hunt, A., & Lester, P. (2019). Novel biotechnologies for eradicating wasps: Seeking Māori studies students' perspectives with Q method. *Kōtuitui: New Zealand Journal of Social Sciences Online, 14*(1), 136–156.

Moeed, A. (2016). Novelty, variety, relevance, challenge, and assessment: How science investigations influence the motivation of year 11 students in New Zealand. *School Science Review, 97*(361), 68–74.

Moeed, A., & Cassidy, S. (2020). *Teaching and learning science in school an indigenous school. A novel approach with a Māori world view.* Azra Moeed.

Nolan, A., & Molla, T. (2018). Teacher professional learning in early childhood education: Insights from a mentoring program. *Early Years, 38*(3), 258–270. https://doi.org/10.1080/09575146.2016.1259212

Palmer, D. H. (2009). Student interest generated during an inquiry skills lesson. *Journal of Research in Science Teaching: The Official Journal of the National Association for Research in Science Teaching, 46*(2), 147–165. https://doi.org/10.1002/tea.20263

Pluim, G., Nazir, J., & Wallace, J. (2020). Curriculum integration and the semicentennial of Basil Bernstein's classification and framing of educational knowledge. *Canadian Journal of Science Mathematics and Technology Education. 20,* 715–735. https://doi.org/10.1007/s42330-021-00135-9

Rameka, L., Ham, R., & Mitchell, L. (2021). Pōwhiri: The ritual of encounter. *Contemporary Issues in Early Childhood,* 1463949121995591.

Ritchie, J. (2005). Implementing Te Whāriki. In *Practical transformations and transformational practices: Globalization, postmodernism, and early childhood education.* Emerald Group Publishing Limited.

Rogoff, B. (2003). *The cultural nature of human development.* Oxford University Press.

Sinnema, C., Daly, A. J., Liou, Y. H., & Rodway, J. (2020). Exploring the communities of learning policy in New Zealand using social network analysis: A case study of leadership, expertise, and networks. *International Journal of Educational Research, 99,* 101492.

Skerrett, M., & Ritchie, J. (2021). Te Rangatiratanga o te Reo: Sovereignty in Indigenous languages in early childhood education in Aotearoa. *Kōtuitui: New Zealand Journal of Social Sciences Online, 16*(2), 250–264. https://doi.org/10.1080/1177083X.2021.1947329

Vellopoulou, A., & Papandreou, M. (2019). Investigating the teacher's roles for the integration of science learning and play in the kindergarten. *Educational Journal of the University of Patras UNESCO Chair.*

Veresov, N. (2017). The concept of perezhivanie in cultural-historical theory: Content and contexts. In *Perezhivanie, emotions and subjectivity* (pp. 47–70). Springer.

Zhang, Q. (2017). Do learning stories tell the whole story of children's learning? A phenomenographic enquiry. *Early Years, 37*(3), 255–267. https://doi.org/10.1080/09575146.2016.115 1403

Chapter 6
Conclusion and Final Thoughts

Findings, and Implications for Policy, and Practice

Abstract In the previous chapter we discussed the emerging themes in the light of recent, relevant literature. Here we will share the answers to the research questions based on our findings and ponder on the complexity of curriculum policy, its implementation, and implications for practice. We began the research with three broad questions: First, about how the mentor teacher interactions might enable science teaching by a non-specialist ECE teacher. Second, the value of teacher inquiry into her own science learning, teaching, and understanding about the nature of science investigation. Third, the benefits of using science investigations to develop children's science ideas and literacy skills. These questions are answered in the following.

Keywords Complexity of policy · Implementation, and practice · Aspirational curriculum needs teacher professional development for aspirational implementation

6.1 Introduction

The research began under the first ECE curriculum Te Whāriki (Ministry of Education, 1996) and concluded with the implementation of the revised Te Whāriki (Ministry of Education, 2017). Relevant to this research is that the emphasis changed from 'Following the child's interest' to promoting 'deliberate acts of teaching'. There was also a shift from reporting progress on goals to setting learning outcomes and making evidence-based evaluation of the fulfilment of these outcomes. Exploration began with setting up activities for children to choose from, the teacher noticing the child's interest and understanding their dispositions and schema.

The shift in emphasis is towards exploration which is purposefully designed with a clear set of learning outcomes. The children are encouraged to develop their working theories and with teacher support and playful learning activities, experiment and refine their theories by problem solving and trial and error. Such a focus is exciting, but teachers will need support and this book is expected to provide some starting points for those teachers who choose to read it. For teacher educators, modelling and providing opportunities to explore alongside their peers would be a useful starting point. Unpacking this approach and following this with thinking about the next learning steps would be a useful way forward.

The other significant curriculum change was children developing a Māori world-view. Here, children learn to understand their world from a Māori perspective. Just as the teacher in this research had to learn science ideas and how to teach them, the teachers will need to develop Mātauranga Māori knowledge. If the intent of the bicultural curriculum is to flourish, then developing this knowledge will be critical. New Zealand is constitutionally a bicultural country where there exists a partner-ship between Māori, the indigenous people of the land and the Pākehā (European) settlers. In addition, the society is increasingly becoming multicultural. The teacher registration standards have specific requirements for a more inclusive education now.

For example, Code of Professional Responsibility and Standards for the Teaching Profession (Education Council, 2017) states, "New Zealand is an increasingly multi-cultural nation, and Te Tiriti o Waitangi is inclusive of today's new settlers" (p. 4). Therefore, addressing the cultural needs of the multicultural society should not in any way reduce importance of the bicultural nature of our country.

6.2   Answers to the Research Questions

This section presents answers to the three research questions drawing upon the evidence presented in earlier chapters. As the focus of the research was on science, we have included only a few literacy and numeracy learning stories in Chap. 4. The literacy and numeracy examples will be explained fully in response to the third research question.

6.2.1   What Teacher and Mentor Interactions Support ECE Teacher's Science Teaching Pedagogy?

Mentoring began with encouragement to try, acknowledging the learning, and guid-ance about the appropriate science ideas for little children to grasp. Then sharing ideas about some activities that would encourage children to play and explore, but also to observe and think. This approach was helpful in building confidence. Clearly, there was a need at this point to make science visible in what the children may be doing anyway, for example, waterplay. Questions like, how about giving children things that they can work out will float or sink? Asking whether a balloon filled with water will float or sink? What about one that is frozen? Pointing out that 'Who sank the boat'? Might be an excellent book to read with the children in this context. The joy that the teacher felt when a child said it was the mouse that sank the boat provided the teacher intrinsic motivation to try other things. A brief discussion about light things floating and heavy things sinking lead to the ideas of mass and density.

By the third year, the teacher was herself challenging the children to make things that would normally sink, float.

At the same time pointing out that in science you experiment, and you do not always have to be right, when things do not work, try again. Science is evidence based when children offer an explanation 'how do you know'? Is the right question to ask. This practice gets the children to look for evidence and provide evidence-based explanations.

In the later years when children played with magnets, the teacher asked why something is pulled towards the magnet, a child said it is made of metal. The teacher asked, will the magnets pull all metals and handed the children her gold chain which was not attracted to magnets. One parent whose child became really interested in this provided samples of other metals for the children to explore. This whānau input was most welcome and is encouraged by the curriculum.

Timely advice, listening to how things worked out, appreciating the learning stories that reported science learning were important. From the mentoring perspective, it was like teaching a child to walk or cycle.

Encourage → Hold hand → Offer trainer wheels → Celebrate independence

Mentor support at critical times to unpack complex changes and helping the teacher find the solutions herself, giving an idea and allowing her to make it her own was a useful approach. For example, the ice balloon activity started out as a floating and sinking activity. However, the teacher produced three balloons, one filled with air, one with water and one with ice. This became an ideal sensory exploration, and the children had the opportunity to talk about how each felt. The air balloon was lighter, the water balloon squishy, and the ice balloon was hard and VERY cold. And then the ice balloon melted! The teacher commented, "that was so good for new vocabulary and literacy"!

The mentoring mantra that worked in this case was to progressively and with support hand over the teaching and learning responsibility to the teacher. Gradually, moving away from telling and showing to listening and giving well-earned praise. This could be likened to finding one's own voice through discovering a vocabulary and using it in a personal, at times idiosyncratic manner.

6.2.2   How Can Teacher Inquiry into Their Practice Build Their Science Knowledge, Knowledge About Science, and Nature of Science Investigation?

Teaching as inquiry was first introduced in The New Zealand Curriculum (Ministry of Education, 2007). It gradually filtered down so that the ECE centres started asking their teachers to inquire into their own practice. This was a shift from earlier focus on teachers reflecting on their practice. From a mentoring perspective, the inquiry cycle had to be unpacked. Why would one do an inquiry and how might it help in planning future learning had to be discussed with examples. The mentor notes show that several examples of possible inquiry questions were discussed.

> What is our centre doing to help children learn te reo Māori? (Question) What would make it more effective? (Method). How will we know? (What evidence to gather analyse and report).

> Which area in our centre does not attract many children to play there? Why might that be? What can we do to make it more attractive? How would we know that it had become more attractive? (Mentor notes, Year 4).

These questions clarified the purpose of doing an inquiry. How would we know? Was about gathering evidence? Some way of measuring? Finally evaluating the changes.

The challenge that the teacher faced was a research project that the centre she was working in agreed to participate in. This should have been helpful, but it turned out that the inquiry process proposed had 10 pages, each was a template to be completed. A valuable lesson here was for the researcher to understand that when starting something it is important to keep the process simple. Metaphorically, a child learns to sit, then crawl, then stand up and walk. The child may at this stage be able to run, but even a half marathon is too ambitious! In brief, these were new beginnings for the teachers. The whole process put a fear of doing any inquiry into the teacher's and her colleague's psyche. Sharing the framework presented in Chap. 3, Fig. 3.2 with a lot of teacher input and thought made the job easier. The framework was inclusive, collegial, and supported by all at the centre including the manager. The butterfly inquiry, which was intended for one summer, continued for three years with repeated cycles. With each passing year more science ideas were included.

The key here was to ask questions, encourage trial and error, extending the exploration when children asked new questions. According to the notes, in Year 5 while continuing the butterfly inquiry, the teacher wrote in a learning story.

> Children counted the number of legs that a caterpillar had in a picture. Then today we had a real caterpillar from our swan plant. Sam was very excited that there were legs on the other side too! Great observation and wonderful counting, Well done.

The teacher was very excited as she saw one other benefit to doing this inquiry. Children also used magnifying lenses and worked out how to use them correctly. When something worked and the teacher had a picture to share, for example, children sorting leaves, it was good to tell her that sorting and identifying is a type of science investigation. Similarly, when children arranged the leaves from the largest to the smallest it was a 'pattern seeking investigation'.

Gradually, the teacher basically used all approaches to science investigations as reported in Chap. 4. The children in her care, made models, used models, did fair testing to find out if their snails liked lettuce or cabbage leaves better. Growing plants from seeds, counting number of leaves to measure plant growth, and learning to take care of their plants gradually became embedded in everyday practice.

Children learnt science ideas of floating and sinking; melting and freezing, pushing, and pulling, using their senses to explore and language to describe.

Teacher interest, willingness to try and mentor support has made this teacher an excellent teacher of science in ECE. She understood the value of trying things out before putting them in front of the children, observe carefully and always thinking what the next learning opportunity would be. She learnt that science is evidence-based, theories can change when new evidence is found and importantly, experiments don't always work, such is the nature of science (Lederman & Abd-El-Khalick, 2002).

6.2.3   In What Ways Can Science Investigations Support Children to Make Sense of the Physical Natural World from a Scientific Perspective?

To answer this question we first looked at what had been learnt and then draw attention to the place of curriculum in ECE.

Perhaps one of the exciting findings was that children experienced all different types of investigations (Goldsworthy & Watson, 1999; 2000; NZC, 2007). In doing so they learnt to make close observations using tools such as a magnifying lens. Learning to make observation is essential scientific practice and all scientific observations are theory laden (Millar, 2005). During their exploration, encouraging children to count the legs of an insect or spider focused the children on learning that theoretically knowing the number of legs helps to identify whether a creature is an insect or a spider. Similarly, guiding the children to use their sense to differentiate between hot and cold; sweet, salty, or sour, helped them to use their senses in decision making and understand that there is a range within the experience of hot and cold. Hirsh-Pasek et al. (2009) proposed that discovery-based, active learning is a powerful pedagogical approach. Following further research Hirsh-Pasek et al. (2015) claimed that learning was enhanced when adults provide learning environments that encourages fun-led exploration and discovery. Such an environment was repeatedly provided by the teacher in this case.

Zosh et al. (2018) argue that guided exploration is more likely to lead to science learning as was the case when the children investigated the magnet. For this to happen the teacher herself needed to learn that scientists investigate in many ways. With experience over time the teacher presented numerous explorations. From recent research in New Zealand Freeman (2021) described provocations as activities that invited children to engage with the materials and the child is attracted to it and is responsive. This was the case with the different and varied exploration the participants engage in.

We acknowledge that the evidence shared is across many years but believe that each new exploration introduced by the teacher, invited children to play, then be guided to make close observations, talk about their current theory, experiment, think and make decisions. By the end of this research the teacher was using all the tools in her tool kit and children were playing and the teacher knew what they were learning.

Wood and Hedges (2016) suggest that struggle over the place of curriculum in ECE reflects the influence of developmental psychology in policy frameworks and approaches to curriculum pedagogy and assessment. They conclude that critical questions about curriculum in ECE are essential to create theoretical frameworks to understand ways to consider, think about curriculum alongside pedagogy, assessment, play, and learning.

Experiencing science and mathematics through play and play through science and mathematics provides authentic spaces, contexts, and conversations between peers and with teachers (Parks, 2015). When play-based programmes are prioritised it is considered that children choose activities based on their interests and the materials available to them. Sometimes, play-based activities are interchangeably considered to be child-centred. McCormick Smith and Chao (2018) argue that although these can be interpreted variously, the key is 'student agency'. In both cases, when teachers utilize children's questions and make meaningful materials and contextualised activities accessible in an ECE centre, it does not automatically mean that the child has lost the agency to choose what interests them nor does the approach necessarily become teacher centred. Inquiry-based learning activities are complex but have substantial potential for encouraging children's knowledge construction; when integrated with technology, this potential is magnified (Wang et al., 2010). The many examples in the present study saw the teacher providing varied, interesting, and age-appropriate materials. Children were not asked to come and take part, neither was a lot of science fact downloaded on to them by the teacher.

Seema's approach was to bring things along and put them out for children to explore. In each activity she ensured that children had plenty of time to play, think and talk. She learnt to listen to them talk about what they were doing before asking a question that allowed them to think deeply about what they were doing. There were many occasions when the children were attracted to an activity, played for a while, and moved on to something else and that was acceptable too. If some seeds, pots, and potting mix were put out it was an invitation to come and feel the texture of the mixture. If children sat down and wondered what these things were about, it was an opportunity to talk and perhaps plant some seeds. Then talk about what care it might need and put their pots out again so they could water, transplant them outside and continue to look after them. We believe, this is not a school activity that the whole class is made to sit and do as the teacher instructed with the teacher talking at them.

The evidence we have presented here shows that the activities, materials, resources put out to attract the children were neither school nor ECE specific. There was water play, messy play, floral play, magnetic play, snail, and insect exploration alongside blocks, painting, drawing, puzzles, books, and outdoor activities. We believe, these put out an open invitation to all to come and play, be interested in, or move on. They were things that the children could explore and play with. The child-teacher talk led to language development and children being able to communicate their ideas and make contributions to everyone else's learning through conversation as required by our curriculum.

It was when the learning stories were analysed that as researchers, we were very excited that these children had experienced all approaches to science investigation.

It was never planned by the teacher, "today we are going to do a fair testing investigation". But if flowers were being put in coloured water, it was an opportunity to put one in without water, so the children could learn that flowers needed water to stay alive. As is clear from the evidence presented, there were certainly more exploration opportunities.

Another exciting finding is that children can begin to make sense of the world in a scientific way and start to understand the nature of science, for example experiencing the many ways in which scientists do science. Developing Nature of Science understandings is being prioritised in ECE teaching and issues concerned with content (Hansson et al., 2020). It is argued that introducing science in early years made present a new way of teaching science in ECE, where children are introduced to discussions about how, why and by whom scientific knowledge is developed.

6.2.4   In What Ways Can Science Investigations Be Integrated into an ECE Learning Programme and Support children's Literacy Learning and Their Holistic Development?

The child's holistic development was a key focus for the original Te Whāriki. In the first years the focus was on belonging, physical, social, and emotional development. Interestingly, it appeared that the emphasis was to provide care for the child to grow up in a healthy environment which acknowledged their culture, particularly Māori culture. This emphasised the three Te Tiriti o Waitangi principles of partnership, protection, and participation. These were not well understood by the teacher. In the beginning partnership was seen as the parents and teachers being responsible for care, protection was almost understood as protecting the child and keeping them safe, and participation as children participating in all activities. There was a gradual move towards whānau and teachers both having input in the children's *care* and *learning* and that Te Reo Māori is a taonga (treasure) which needs to be protected. This would happen if all children learnt to speak the language. Finally, participation was about all children having equal access and opportunity to participate in all activities. The emphasis was inclusion of Māori children, who were not well represented in the ECE centres.

As the teacher created more and more opportunities for the children to investigate, children were learning science specific vocabulary.

> The snail had a shell, when there is danger, it goes in the shell. Snails come out at night; they are nocturnal. They have tentacles to feel and eyes to see (Year 5 story).

The next step was to encourage children to speak in sentences rather than giving one-word answers. Oral language was also encouraged by asking children to talk about what they were doing.

In year 6 of the study, concerns were being raised about the low level of literacy and numeracy among children in primary schools. This drew attention to the communication strand of the curriculum and that the ECE centre should support children to have better literacy and numeracy when they started school. Seema was given charge of the four-and-a-half-year-old children as the 'transition to school group'. Before the children left, they were all able to count to 20 and some to 100. They were able to group counters into sets of 2, 3, 4 and 5 and then count the lot. In her current ECE centre, through playing board games such as snakes and ladders as well as using number puzzles children are developing their numeracy skills and mathematical thinking. All children can not only recognise shapes but name them and find objects of those shapes in amongst their toys. Interestingly, Seema noted that one child picked up a ball and said, "This is *not* a circle". She explained that it was a sphere and why.

As the children learnt the names of shapes, she combined this with naming the colours in both English and Te Reo Māori. Some children were able to count backwards. Later, she encouraged children to count in Te Reo Māori. She collaborated with the parents to find out information so that the children could learn their mihimihi. This is how Māori introduce themselves in their language. All children in the transition group could orally recite their mihimihi with confidence. A typical mihi might be:

In Te Reo Māori	Meaning in English
Ko – – – – – – – – – tōku maunga	My mountain is…
Ko – – – – – – – – – tōku awa	My river is…
Nō – – – – – – – – – ahau	I am from……
Ko – – – – – – – – – tōku kura	My school is…
Ko– – – – – – – – – tōku matua	My father is…
Ko – – – – – – – – – tōku whāea	My mother is……
Ko – – – – – – – – – tōku ingoa	My name is…

Learning the mihimihi, helps children to develop their identity, make links with the local geology, beginning to think about who they are and where they stand. Whakapapa (Genealogy) is very important to Māori children and would support all children to consider their relationship with the natural environment.

In the last year of this research using the strategies promoted by A Ministry of Education resource Te kōrerorero (**Talking Together, Te Kōrerorero—Education Gazette**), the teacher began to deliberately follow the process set out in Fig. 6.1 for oral language development. She put a copy of this poster on the wall in their classroom so that it was a reminder to her and useful to her colleagues.

She set herself the goal to use at least two of these strategies each week. This was reflected in the learning stories and learning conversations written by her. She also used photographs of a street scenes to help children verbalise the visual symbols in everyday life, for example, Traffic lights, symbols for pedestrian crossings, schools,

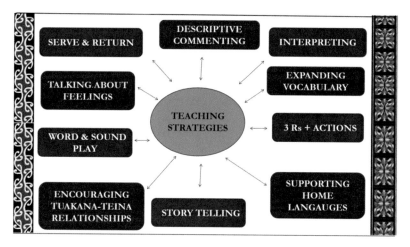

Fig. 6.1 Teaching strategies for oral language development

bus stop, and other traffic signs such as Give way and STOP. Seema made this into a game where children could spot a symbol and say what it was.

In our country, we have three national languages, English, Te Reo Māori, and Sign language. Children can sing several actions songs in the first two languages, they can identify colours, shapes, and are beginning to understand and speak instructions in the two spoken languages. In the final year of this research, she began teaching sign language to the children.

Children use language to communicate with each other. We may think of language as words, sentences, or stories but there are other languages of sign, mathematics, and creative ways of communicating, for example, through art, music, and dance. We found that as children learnt science, they not only learnt mathematical concepts of patterns but measuring and comparing sizes and organising patterns when completing puzzles. Learning Science supported the children to meet the intended learning outcome of "Recognising mathematical symbols and concepts and using them with enjoyment, meaning and purpose" (Ministry of Education, 2017, p. 42).

6.3 Complexity of Policy, Implementation, and Practice

As said earlier, Curricula by their nature are aspirational and so they ought to be. However, they should not be unachievable for most teachers. The quality of care, teaching and learning depends on the teaching workforce having a clear understanding of the vision, goals, and intentions of the curriculum. We have observed three major changes in New Zealand Education in the last two decades. The introduction of the National Certificate in Educational Achievement (NCEA) that replaced the assessment approach in secondary schooling. Introduction of the New Zealand

Curriculum (Ministry of Education, 2007) and relevant to this research the updating of Te Whāriki (Ministry of Education, 2017). In each case the government of the time invested huge amount of money in setting up the NCEA framework, and the writing of these curricula. What is surprising is that when these brilliant framework and glossy curricula are published there was hardly any funding left to make a significant investment in teacher professional development to see them to fruition. If the policy makers believe that the aspirational goals, they have published are worthy of implementation, funding needs to be divided into two lots at the very start. One pot for the policy makers to come up with the masterpiece and IMPORTANTLY, the second pot for teachers to be helped to upskill, learn, try out, critique their own practice with support. Implicated in the last should be significant efforts to ensure ongoing evaluation in preparation for what will be a new wave of reform sometime in the future.

History tells us that after many years of implementation, research reports that 'teachers have not implemented the curriculum as intended'. Or that "authors have used innovative schools to illustrate how an assessment approach that honours the intentions of the curriculum can provide rich learning experiences that motivate students and deepen their learning experience" (Hipkins et al., 2016). We point out that a few innovative schools are not **most** New Zealand schools. Further that "teachers have not been well-supported to make the shifts in curriculum thinking" (Hipkins et al., 2016, p. 206). It is remarked that, 'the poor dears did not understand the intentions!' Why not, because the effort made for them to understand the intention was minimal.

Returning to the case of ECE, not all the teachers are trained and there are few teachers who stay in one centre for long enough to share their wisdom with their colleagues. The current practice of uploading more and more documents and resources on the Ministry of Education website is not enabling. It does not consider that the ECE teachers are on their feet for 7 or 8 hours a day. They start the day by setting up in the morning, supervise the children and care for them then, write learning stories, maintain children's portfolios, build relationships with the children's whānau, tidy up at the end of the day and attend a meeting at least once a week after work. Where is the time to locate, think, use, and reflect on the excellent resources on the website? ECE teachers do not have school holidays which are considered non-contact time for other sectors to do professional reading and learning.

The findings of this research have shown that change is possible when learning opportunities are accessible. Science can be taught as children play in their ECE centres.

6.4 Final Thoughts

We have in this book reported the progress and learning of a teacher over a long period of time and the benefits it has had for the children's learning. The mentoring was not time consuming but timely help. We acknowledge that the mentor having been a science teacher and science teacher educator with experience in mentoring pre-service and in-service teachers was helpful. Her willingness to delve into and develop a nuanced understanding of the ECE curriculum and the proposed changes during this research helped the mentor to see science teaching and learning from early childhood perspective. Teacher educators and science teachers with similar experience of mentoring, if interested in ECE science teaching and learning have the capacity for similar mentoring.

This could be a good approach to in-service teacher professional development, as opposed to shorter courses of perhaps only a few hours. Mentoring has the advantage of being paced to the needs of the recipient and offered when required. It is like seaweed drifting in on the current, offering nutrients, then retreating to replenish and return. But even this metaphor fails to capture how the mentor has gained much from the ECE teacher they have mentored. It is a two-way relationship based upon the strength of the inter-relationship.

From the teaching science and science investigation perspective, we do not consider the fact that all ECE teachers who have gone through the New Zealand education system have learnt science until at least Year 10, most up to Year 11. Therefore, to say that they do not have the science background to teach science can only mean three things. One, that the children's science learning to the age of 14–15 years has no value; it has been forgotten. Two, that we do not believe what they had learnt earlier in life can be relearnt with encouragement and a little support. Three, that it is an easy default position to say I know no science or 'they do not have a science background', the later responses we believe are excuses.

This book in the hand of any policy maker, teacher educator or teacher will show that you only need to understand the science you were taught in school to teach it in ECE if you had good secondary education.

Finally, in the words of President John F. Kennedy worthy of further reflection in our context:.

One person can make a difference, and everyone should try.

Perhaps, this should be the underpinning theory for those who write the policies. Allow funding to support teachers to translate the *aspirational curriculum* into *aspirational teaching and learning* process. It is possible, and will undoubtedly require greater investment.

References

Education Council. (2017). *Our code our standards code of professional responsibility and standards for the teaching profession.* Retrieved December 14, 2022 from www.educationcouncil.org.nz

Freeman, S. (2021). Provoking opportunities for science in early childhood education. *Early Childhood Folio, 25*(2), 31–35.

Hansson, L., Leden, L., & Thulin, S. (2020). Book talks as an approach to nature of science teaching in early childhood education. *International Journal of Science Education, 42*(12), 2095–2111.

Hipkins, R., Johnston, M., & Sheehan, M. (2016). *NCEA in context*. NZCER Press.

Hirsh-Pasek, K. (2009). A mandate for playful learning in preschool: Applying the scientific evidence. New York: Oxford University Press.

Hirsh-Pasek, K., Adamson, L. B., Bakeman, R., Owen, M. T., Golinkoff, R. M., Pace, A., & Suma, K. (2015). The contribution of early communication quality to low-income children's language success. *Psychological science, 26*(7), 1071–1083.

Lederman, N., & Abd-El-Khalick, F. (2002). Avoiding de-natured science: Activities that promote understandings of the nature of science. In *The Nature of science in science education: Rationales and strategies* (pp. 83–126).

McCormick Smith, M., & Chao, T. (2018). Critical science and mathematics early childhood education: Theorizing reggio, play, and critical pedagogy into an actionable cycle. *Education Sciences, 8*(4), 162.

Millar, R. (2005). What is 'scientific method' and can it be taught? In *Teaching science* (pp. 172–185). Routledge.

Ministry of Education. (1996). *Te Whāriki: He Whāriki Mātauranga mō ngā mokopuna*. Learning Media.

Ministry of Education. (2007). *The New Zealand curriculum*. Learning Media.

Ministry of Education. (2017). *Te Whāriki. He whāriki Mātauranga mō ngā mokopuna o Aotearoa: Early childhood curriculum*. Retrieved from https://www.education.govt.nz/assets/Documents/Early-Childhood/ELS-Te-Whariki-Early-Childhood-Curriculum-ENG-Web.pdf

Parks, A. N. (2015). *Exploring mathematics through play in the early childhood classroom*. Teachers College Press.

Wang, F., Kinzie, M. B., McGuire, P., & Pan, E. (2010). Applying technology to inquiry-based learning in early childhood education. *Early Childhood Education Journal, 37*, 381–389.

Watson, J. R., Goldsworthy, A., & Wood-Robinson, V. (1999). Practical science investigations in England and Wales: A national survey. In *ESERA conference, Kiel*.

Watson, R., Goldsworthy, A., & Wood-Robinson, V. (2000). 8 SC1. *Issues in Science Teaching, 70*.

Wood, E., & Hedges, H. (2016). Curriculum in early childhood education: Critical questions about content, coherence, and control. *The Curriculum Journal, 27*(3), 387–405.

Zosh, J. M., Hirsh-Pasek, K., Hopkins, E. J., Jensen, H., Liu, C., Neale, D., & Whitebread, D. (2018). Accessing the inaccessible: Redefining play as a spectrum. *Frontiers in psychology, 9*, 1124.

Glossary

Te Reo Māori words	English translation
Ako	Reciprocity of teaching and learning
Ākonga	Learner
Kai	Food
Kaiako	Teacher
Karakia	Prayer
Kura	School
Mahi	Work
Tamariki	Children
Tangata	People
Tangata whenua	People of the land, referred to the 'first inhabitants'
Te Kōhanga reo	Māori immersion early childhood education
Te Reo	Language, voice
Te Reo Māori	The Māori language
Te Whāriki	The early childhood curriculum
Te Whāriki	A woven mat (metaphor for ECE curriculum)
Tikanga	Customs, ways of doing
Tūpuna	Ancestors
Whānau	Family

A. Moeed et al., *Playful Science Investigations in Early Childhood*,
SpringerBriefs in Education, https://doi.org/10.1007/978-981-99-7286-9

Printed in the United States
by Baker & Taylor Publisher Services